BLUEJACKET

Seaman to Captain (in a mere 30 years)

DAVID FRANCIS
Captain, USNR (Retired)

BLUEJACKET

Seaman to Captain (in a mere 30 years)

DAVID FRANCIS
Captain, USNR (Retired)

RED BIKE PUBLISHING, LLC

Bluejacket
Published By Red Bike Publishing
www.Redbikepublishing.com

ISBN: 978-1-936800-36-0

LCCN: 2022931748

TABLE OF CONTENTS

PROLOGUE AND DEDICATION

"Honor to the Soldier, and Sailor everywhere, who bravely bears his country's cause. Honor also to the citizen who cares for his brother in the field, and serves, as he best can, the same cause— honor to him, only less than to him, who braves, for the common good, the storms of heaven and the storms of battle."
—Abraham Lincoln, letter to George Opdyke and others, December 2, 1863

"Thank you for your service." I hear that greeting more often than I used to. Whenever someone thanks me for my service, I often happily respond with, "You're welcome. The Navy been berry, berry good to me," mimicking, of course, the voice of Chico Escuela, a character on *Saturday Night Live* portrayed by Garrett Morris.[1]

If you served in the military, married into it, gave birth to a service member, or in any manner supported one, you earned the right to question, complain, and gripe; and you also had the right to be proud. Thank you to those, in and out of uniform, who bore the burden and won the peace; for those who returned to homecomings and those who did not; and for their families, loved ones, spouses, and friends who watched, waited, endured, and supported. Thank you for your service and sacrifice. I dedicate *Bluejacket* to you.

Those in my generation were called "war babies," born during the Second World War (WWII). We "war babies" came of age in the 1950s and '60s during the height of the Cold War.

---

[1] Chico Escuela was a fictional baseball player from the Dominican Republic who spoke limited and halting English, and he responded to most any interview question with his catch phrase: "Baseball been berry, berry good to me."

Remember the Cold War? For many years, for most of my lifetime of naval service, the Western democracies faced off against the Soviet Union and the communist bloc countries competing for world dominance. The Cold War was a long, costly, and frequently frightening military competition with the former Soviet Union. At times, our survival was in doubt.

I proudly wore Navy Blue for thirty-three years—from the building of the Berlin Wall and the Cuban Missile Crisis to the collapse of the Soviet Union. Since my birth, the United States has fought many, way too many, wars. We have continually been called to service and sacrifice in a Cold War, the War Against Terror, on battlefields in Korea, Vietnam, Iraq, Afghanistan, Kuwait, Africa, (I may have missed some) and at home.

In the midst of the ravages of a global pandemic and murderous terrorism, memories of the Cold War are faded. As a US submariner, I played a small role in defending our freedom and survival. I write this book also to honor my fellow submarine sailors who spent days, weeks, months, and years from home, tolerating loneliness, isolation, and hardship to provide the nuclear shield and our hidden line of defense.

At sea for decades were United States diesel electric, nuclear fast attacks, nuclear missile submarines, and sailors who remained hidden beneath the waves. Inside their black, submerged hulls, many carried thermonuclear missiles with more destructive firepower than all of the bombs dropped during World War Two, including the two atomic bombs, as we were often told. It was more than just a global chess match. There were countless Cold War casualties: victims, deaths, and immense physical and emotional damage.

My thanks and gratitude to the millions of Americans, both military members and civilians, who fought and won the Cold War. Second only to our still-unfinished battle for equal rights, justice, and freedom, it is America's longest struggle, lasting well over forty years. Those decades of stress were a crucible, a severe test and trial, and a defining experience for me and for a great many others. That's why I wrote *Bluejacket.*

As a nuclear missile submarine officer during the Cold War, I was in the thick of it. I was part of the action. And my Navy was the primary reason we survived and ultimately won. You may also remember that, for many weeks, we found ourselves at the brink of war during the Cuban Missile Crisis in the fall of 1962. We survived, I contend, because of our sea power. We prevailed because we controlled the seas.

US President John F. Kennedy told us as much on June 6, 1963, on board the USS *Kitty Hawk*: "Events of October 1962 indicated, as they had all through history, that control of the sea means security. Control of the seas can mean peace. Control of the seas can mean victory. The United States must control the seas if it is to protect your security. . ."

This, my sea story, is also a love story—an account of learning and surviving; a narrative about the grandson of Irish immigrants who, raised in scarcity and disadvantage, achieved the "American dream" of financial security and social mobility. I choose the term "scarcity" instead of "poverty," but we were poor. But among my good fortunes along the way, I met my love early in my life, and it has lasted the test of time and trial. I frequently call it my rags-to-riches story. My wife Linda is still waiting for the riches.

A lot can happen to a sailor during thirty-three years of naval service. And it did. Many people taught me many things. In the course of my writing, I returned to those events when love, learning, resilience, human compassion, gratitude, and grit helped me get through some tough times and build a meaningful life and career.

I hope that my account of overcoming disappointment, hardship, and adversity, and my attitude of gratitude for the love, encouragement, and support, from many people, will be helpful to others navigating their own unique course in life.

It was my privilege to serve and advance from US Navy seaman recruit to US Navy captain in a mere thirty years. I owe my education and social advancement to the United States Navy and to all Americans. The Department of the Navy, and my fellow taxpayers, financed both my undergraduate and graduate education and opened a path for a poor kid from Michigan to serve as a nuclear submarine officer and then retire as a senior naval officer.

In *Bluejacket,* I describe my journey: the ups and the downs (an old submarine joke), the struggles, the achievements, the loneliness, the love, and the sweet homecomings.

Prepare to dive.

CHAPTER ONE

A BOY WITH A DREAM

Navy flag, courtesy Naval History and Heritage Command (NHHC) catalog #KN-471.

My dream was to earn a scholarship to the US Naval Academy at Annapolis, Maryland, because that's where naval officers' careers get their best start.

The United States Naval Academy is a federal service academy adjacent to Annapolis, Maryland. The campus is located on the former grounds of Fort Severn at the confluence of the Severn River and Chesapeake Bay. Established in 1845 under Secretary of the Navy George Bancroft, it is the second-oldest of the five US service academies, and educates officers for commissioning primarily into the United States Navy and United States Marine Corps.

I had above-average grades, scored high on the college entrance exams, and received a letter of recommendation from my congressman. That's how entrance to military academies is decided—by congressional appointment.

My first disappointment was that I was the first alternate, not the principal candidate. Being number two on the list meant that, if the principal candidate was accepted, I would not be.

Road trip

Waking up much too early in the morning, my dad and I set out on a Greyhound Bus to the Naval Base at Great Lakes, Illinois, for my Naval Academy physical examination.

The bus was hot and smelly, and I was not allowed to open a window. Annoyed and anxious, I just wanted to get there. We stopped frequently and sat in far too many small bus stations.

Are we there yet?

We spent the night in a small, inexpensive motel near the Great Lakes Naval Station. The next morning, tired and hungry, I reported. Poor planning. To check our physical fitness, we were ordered through calisthenics: pushups, pullups, knee bends and more. The sixteen-count burpees we were required to perform caused pain in my injured shoulder, which I had hurt on a wrestling mat a few months before.[2] I was close to passing out, but I managed to stay alert and on count.

Next was the weigh-in and measurement, like a boxer before a fight. My weight was 106 pounds. Yes. Only 106 pounds.

---

[2] Burpee is a conditioning exercise that includes squats (placing palms of hands on the floor in front of feet, jumping back into push-up position) push-ups, returns to the squat position, and jumps into the air with arms extended overhead.

The minimum height to attend Annapolis was determined to be five foot, four inches. My measurement was five foot, three and a half inches. I was too short!

I was too stunned to be angry, and too young to protest the result. The idea of a lawsuit against the US Navy was farfetched back then. The principal candidate also failed his physical. The second alternate (the one behind me), also from my high school, was admitted to the Naval Academy that year.

On the bus trip back home to Bay City, Michigan, I was sad, subdued, and silent. It was not a father-son bonding road trip; it was a grief trip. I felt despair and rejection. I did not understand it, and did not know how to cope with the loss of my dream and my deep sense of failure.

Since I had placed all my hopes for a higher education on acceptance at the US Naval Academy, and my parents lacked the means to send me to college, I had not applied to any colleges or universities—except for Delta College. Delta was a new community college near Bay City that featured what was described as an expansive and experimental education.

Delta was not Ivy League or even a Big Ten school. But it was a highly affordable junior college. The tuition was only $100 per semester. And it was a lifeline. I was not ashamed to be a junior college student; I felt fortunate. The curriculum was excellent. I received a superb liberal arts education. Eagerly, I studied art, psychology, sociology, music, philosophy, world history, and political science. It was my good fortune to be among a small percentage of human beings who have had the opportunity to attend college and earn sixty college credits on such a budget.

While serving aboard several ships and submarines, I searched for a logical reason why The United States Naval Academy had established that original height "standard." My conclusion was, it's not a physical standard or necessity, but a uniformity issue, an artificial barrier. They didn't want their officers to appear short. Many years later, when Annapolis went coed and admitted women, they were obliged to lower the height limit.

This was the 1960s, and we were constantly aware of the Cold War with the Soviet Union. And yes, in grade school, we actually did the duck-and-cover drill—crawling under our school desks. At that age, I actually believed I would be safe from the nuclear blast.

Yet surprisingly, many evenings I watched *Victory at Sea* and the *Silent Service* on our black-and-white TV. While we lived in fear of nuclear war and annihilation, many Americans often relived our military victories and our national power gained during "The War." When people talked about "The War" they meant the Second World War (WWII).

As a young man, I was inspired by the film depictions of bravery and sacrifice, particularly submarine warfare in the Pacific against the Imperial Japanese Navy. I also admired my Uncle Ned who was a US Navy chief petty officer stationed in Norfolk, Virginia. Ned was a combat veteran of World War II. He served aboard a cruiser during the epic Battle of the Coral Sea and survived his military service.

The Great Lakes

Bay City, by the inland sea, is where I grew up on the shore of Lake Huron. The Great Lakes surrounded me. To our north were the straits of Mackinac where Huron meets Lake Michigan. Above Michigan's upper peninsula was the big lake they called *Gitche Gumee,* Lake Superior, with its countless lighthouses, lakes, bays, islands, and rivers—Great Waters as far as the eye can see. Michigan, from the Ojibwa word for large lake (*Michi Gami*), is surrounded by four of the five Great Lakes: Superior, Huron, Erie, and Michigan. Only Lake Ontario doesn't touch Michigan.

It's not the seacoast, but the Great Lakes in the heartland of America have many miles of sandy coastline and rocky shores, and together they supply a fifth of the world's supply of freshwater. The Great Lakes are my part of America, which shaped Midwestern culture and attitudes. I was shaped by them; they inspired me. Years later I discovered that many sailors in the US Navy had enlisted from my part of the country.

My hometown has a long maritime history. Several Michigan ship builders had constructed wooden and steel vessels, pleasure craft, and warships. President Kennedy's presidential yacht was built in Bay City by Defoe Shipbuilding Company (Originally the Defoe Boat & Motor Works when founded in 1905). Over their nearly seventy years in business, in addition to yachts, Defoe Shipbuilding built torpedo/submarine chasers, minelayers, patrol craft, tugboats, destroyers and destroyer escorts, icebreakers, and landing craft. Of course having a Navy ship builder in town brings in sailors. My wife Linda's sister met and married one.

Our US Navy conducted annual Great Lakes tours. Though done largely for recruiting purposes, it was also a great opportunity for Midwest civilians like me to see an active US Navy vessel. One year a submarine, an updated World War II diesel-electric, steamed up the Saginaw River and moored in Bay City. My sister, who had her driver's license, agreed to take me to the pier. We were thrilled to board the submarine and go down the hatch. The sailors alternated between pointing out the ship's components to me and flirting with my sister.

It was smelly and cramped, but I was hooked. It further motivated me to serve in "The Silent Service."

What did I learn?

- ✓ Always have one more goal!
- ✓ Chart your own course in life. Don't leave it to chance, or to others.
- ✓ Take the helm.
- ✓ Remember the advice of Dr. Suess about "life's great balancing act" in *Oh, the Places You'll Go.*

CHAPTER 2
YOU'RE IN THE NAVY NOW

"Any man who may be asked in this century, what he did to make his life worthwhile, can respond with a good deal of pride and satisfaction, "I served in the United States Navy." —President John F. Kennedy, *Public Papers of the Presidents*, August 1, 1963.

Enlisted Man

Elected in 1960, President John F. Kennedy inspired many of my generation to public service, including me. The Peace Corps, launched by the Kennedy Administration, gave Americans an opportunity to serve humanity overseas. Kennedy also challenged us to fight against the "communist threat." John Kennedy was one of my first boyhood heroes.

To me John Kennedy embodied commitment and courage. Commitment to public service and courage to make the world a better place. President Kennedy's inaugural address on January 20, 1961, spoke to me. I was inspired and determined to help "bear any burden and meet any hardship" to assure our liberty and survival.

"Let the word go forth from this time and place, to friend and foe alike, that the torch has been passed to a new generation of Americans—born in this century, tempered by war, disciplined by a hard and bitter peace, proud of our ancient heritage—and unwilling to witness or permit the slow undoing of those human rights to which this nation has always been committed, and to which we are committed today at home and around the world. Let every nation know, whether it wishes us well or ill, that we shall pay any price, bear any burden, meet any hardship, support any friend, oppose any foe, to assure the survival and the success of liberty."

John Kennedy was also a Navy man with service during WWII. He was a patrol torpedo boat captain. His naval service only added to his appeal to me as a role model.

While JFK was in command of PT-109, in Blackett Strait, south of Kolombangara in the Solomon Islands, on the night patrol, August 1, 1943, his boat was rammed by a Japanese destroyer. Lieutenant Kennedy heroically saved many of his crew and got them safely back to base. Robert Donovan's book, *PT 109* in 1961, and the 1963 movie *PT 109* starring Cliff Robertson as JFK, documented his heroic actions.

Kennedy turned forty-four years old on May 29, 1961, the day I enlisted in the US Navy. Seventeen years old and still in high school, I was required to obtain my parents' written permission to enlist. Perhaps enlisting in the US Navy would give me an opportunity to go to sea aboard a submarine. And I wanted to help win the Cold War.

On the evening of May 29, I arrived at the Naval Reserve Center in Bay City, filled out my enlistment application, and took multiple-choice quizzes. (Later in my career, I learned that my test scores were very high.) Next I met with a Navy doctor for a physical examination. The doctor asked me, "Are you tall enough for this man's Navy?"

Here we go again. Another damn short reference!

Holding my tongue, I passed my physical, and took the oath of office:

"I, David Francis, do solemnly swear (or affirm) that I will support and defend the Constitution of the United States against all enemies, foreign and domestic; that I will bear true faith and allegiance to the same; and that I will obey the orders of the President of the United States and the orders of the officers appointed over me, according to regulations and the Uniform Code of Military Justice. (So help me God)."

Finally I signed my two-year enlistment papers, earning the title of Seaman Recruit Francis. Service members and their "dependents" will recognize that a seaman recruit is an E1 on the military pay scale—rock bottom on the military food chain.

Later I was issued a full set of enlisted uniforms: jumpers and bell bottoms, whites and blues, black shoes, dungarees (the work uniform), the prized navy-blue peacoat (bluejacket), several distinctive white hats, and a brown canvas seabag. My boot camp was not scheduled until August, so I had time to get my trousers shortened and my one Navy seaman recruit stripe sewn onto my left sleeves.

Further, I received critical instructions on how to properly tie a square knot in my black silk neckerchief, and how to form my white hats into the more-attractive, and jauntier, curved shape.

The Republic of South Vietnam

During the turbulent 1960s, the Kennedy Administration was struggling with what to do about the threat to South Vietnam from aggression from communist North Vietnam. In 1961 the US only had about 3,000 military personnel, mostly advisors, in South

Vietnam. However, senior American military leaders believed in the domino theory, the persistent propaganda from the Pentagon. If South Vietnam fell to the communists, the rest of Southeast Asia (Cambodia, Burma, Laos, and Thailand) would fall to communism, like dominoes.

While a mere high school senior, I held the firm belief that the US should not put ourselves in the middle of a civil war in Vietnam. In 1961 most Americans probably did not know where Vietnam was located on a map of the earth. But that would change. Now that a Selective Service Draft Law was in effect, young American men became increasingly concerned and anxious about being drafted into the military and sent to Vietnam.

I saw military service as a necessary and important act of patriotism. I sensed a call to duty to defend our freedom, ensure our survival, and win the Cold War. As a college student and a member of the US Naval Reserve, I knew that I would not be drafted. On my reserve drill evenings, I frequently marched, stood at inspection, and sat in a classroom studying.

The first book issued to me, *The Bluejackets' Manual*, page vii read, "To All Navy Men. You are a Navy man, part of the largest and strongest seagoing force in the world. When you were sworn in and put on your uniform for the first time, you became part of a great tradition . . ."[3] My next study book, *Basic Military Requirements,*

---

[3] *The Bluejackets' Manual*, Fifteenth Edition, United States Naval Institute, Annapolis, Maryland. 1959.

stated, "This book is a self-study training course for men and women in the first three pay grades. . ."[4]

Eager to learn about our Navy, its history, and traditions, I finished several required training manuals, took the tests, then moved on to the next subject—but with one major difference: as I completed my requirements, I could advance in grade and increase my pay.

Great Lakes Naval Training Center (Boot Camp)

As a naval reservist, my boot camp lasted only two weeks rather than eight weeks for "regular" sailors. Reservists learn military knowledge and etiquette, those "knowledge factors," at the reserve center.

My first set of official orders arrived along with my travel vouchers, and I packed my neatly pressed unforms into my indestructible seabag. (It is indestructible; I still have it!) It was my first-ever airplane flight. My destination again was the US Naval Station, Great Lakes, Illinois.

Opened in 1911, Naval Station Great Lakes is the Navy's largest training installation, located on 1,600+ acres, overlooking Lake Michigan.

The propeller-driven Republic Airlines DC3 flew across Lake Michigan in a storm, below the clouds. During that rough, turbulent ride, I got sick to my stomach. Enduring many painful minutes reminded me that I suffered from motion sickness and had decided

---

4 *Basic Military Requirements,* (NAVPERS 10054), Bureau of Naval Personnel, Washington, DC, 1957.

to “go down to the sea in ships . . . and do business on great waters.” (Psalms 107:23, NKJV)

Yup. I had just joined the US Navy.

As we bumped along, rising up, then abruptly falling down in the air, I thought back to my childhood days getting motion sick in the back seat of my parents’ car. And the many carnival rides I went on, knowing I would likely throw up and get kicked off by an angry carnival ride operator.

Finally we landed. But my neatly packed seabag did not land with me. The airline lost it. For three days, before my seabag was returned to me, I wore the same white uniform, washed it in the evening and hung it out to dry. Apparently there were no clothes pins at boot camp. I was given small strings and required to neatly tie the jumper and bell bottoms on the clothesline using little square knots. In the early morning, I put on my clean, white, wet, cold uniform.

My lost seabag appeared magically one evening next to my bunk in the barracks. I don’t know who dropped it off, and I never had the opportunity to offer my eternal thanks.

I had never shaved in my life. And I didn’t think I needed to. However, our drill instructor (DI) disagreed and ordered me to shave what little hairs I had on my face.

As directed, I went to the Navy Exchange (USN’s answer to Walmart) and purchased a Gillette razor with twin blue blades. Each morning I cut my face with the razor in several areas and went through the day, wearing small pieces of toilet paper over each wound.

For the first time in my life, I was given a rifle. I didn't own one, had never fired one, and had never gone hunting. However, my mother was a hunter. She owned a 12-gauge shotgun. And one year, she brought home a deer. She was the youngest of four daughters, and I guess she was "the boy" my grandfather, Arnold, always wanted. Arnold, an avid Michigan hunter, took her out bird and deer hunting. One summer, Arnold made me an accomplice. At night he took me out in a boat on a lake. My assignment was to hold the flashlight and shine the beam on defenseless frogs as he shot them with a .22 caliber rifle, often shooting the rifle directly over my head.

Rifle, Caliber .30 M1

The M1 rifle we were issued was described as "a lightweight, semi-automatic carbine" that was a standard firearm for the US military during World War II, the Korean War, and well into the Vietnam War. We watched an old War Department movie on the care and operation of the M1, then we each took one apart and put it back together.

Every day, I carried it on my shoulder and took it to the firing range. Anyone who has fired the M1 knows that, when you pull back the bolt to load the ammunition clip into the breach, you need to remove your thumb quickly before it gets severely pinched when you release the bolt. The art is to avoid "M1 thumb." Despite being a novice with firearms, I did enjoy daily marksmanship training and never suffered from M1 thumb.

At muster one morning, our DI shouted a question, "Do any of you dummies know how to type?"

I shouted, "Yes Sir!"

"What's your name?"

"Francis, Sir."

"Okay, Francis. You go to the base office. The rest of you ladies march to the galley (mess hall) for mess duty." I was never so grateful to my sister who encouraged me in high school to take a typing class.

For the rest of my brief time in boot camp, I worked part-time in the base office. The Navy chief yeoman (office supervisor) leaned over my shoulder occasionally and said, "Francis, you are really slow."

Each time I turned around and cheerfully responded, "Yes, Chief. But I'm accurate."

Boot camp provides several physical challenges. But at seventeen years old, I was in good shape, because I ran, walked, and biked long distances most of my young life. Having spent many hours in a pool, river, or lake in Michigan, I was also a strong swimmer. So the required obstacle course was actually enjoyable. It consisted of the standard crawl in the mud, climb a high wooden wall, then climb a rope wall. The finale was the hand-over-hand crossing on a rope over a muddy pond. Being a smart-ass kid, after I finished the course, I approached the DI and asked if I could run it again. In a loud voice he shouted, "Get out of my face!"

In submarine school, we were required to simulate in the Escape Tower, an escape from a sunken submarine. In boot camp, we jumped from a high platform into a pool to train for abandoning ship from a surface ship. We wore large, bulky, orange life jackets, which we were warned multiple times to hold onto as we hit the water, or else "it will take your head off." And hit the water feet first, not face first.

It sounds like a joke, but the reason sailors wear bell bottom trousers is so they can take them off quickly. That is, take them off quickly in the water. Which we did in the pool.

Then we tied a knot in each leg, flipped them over our heads to fill the trousers with air, and used them as water wings. After floating around in the cold water for several minutes, we were ordered to get out of the pool. Got that one done.

Each evening in the darkness, I felt an intense sense of homesickness. Other than summer camp, this was my longest time away from home. It wasn't so much the stress of military life, but the absence of loved ones and familiar faces and places.

The Berlin Wall

Among the many planned deprivations in boot camp was the absence of news from the radio, television, or newspapers in the barracks. However word filtered into the barracks somehow that East Germany was constructing a wall between the east and west sides of Berlin. Also a news flash came through that China had bombed Berlin.

As a young student of the Cold War, I believed the wall story, but discounted the China thing. Post WWII, Germany was divided into two countries with Soviet occupation in East Germany, and the

Western Allies (US, Britain, and France) exercising influence in West Germany. Berlin was also divided in half, east and west. To make the Cold War even more dangerous, Berlin was situated in East Germany, and the allies had an agreement with the east side that allowed western access through a corridor into West Berlin.

This situation was a constant source of conflict fanned by the Soviet Union. During the summer of 1961, about 1,000 German refugees had been crossing from East to West Berlin every day. By mid-August, the Soviet Union backed the effort by the communist East German government (German Democratic Republic/GDR) to build the Berlin wall, dividing East and West Berlin.

Fortunately, war was averted, but our anxiety remained high, because we suffered from news deprivation. My office work did not exempt me from learning other important life skills such as swabbing the deck (mopping the floor) and cleaning the head (scrubbing toilets).

I survived the marching, passing in review, the dreaded obstacle course, target practice, the war scare, and my morning shaves. I graduated with pride, looking forward to going home.

What did I learn?

✓ Heroes and historic events can inspire and motivate you.

✓ Find them. Read about them. Learn from them.

✓ A strong US naval presence is needed for our national defense and also for projection of power to keep the peace. Many US sailors are proud to say, "Rest easy, you are being protected by the most powerful Navy the world has ever seen."

CHAPTER 3
THE COLD WAR

USS *Nautilus*, watercolor by Albert K. Murray, courtesy NHHC, C. 1957 Accession #88-195-HL.

Soon after our victory in WWII, Americans began to live with the knowledge and anxiety that we were now in a global economic and military competition with the Soviet Union, our former ally. This "Cold War" included an ever-present fear of nuclear attack. This was my war.

On October 4, 1957, the Soviet Union launched Sputnik (Russian for "traveler"), the world's first artificial satellite and the first man-made object to be placed into the Earth's orbit. This defining event heightened our fears and added the Space Race to our national distress and unease. I was a new, young, cold warrior, joining the millions of Americans committed to fighting and winning the Arms Race and now the Space Race.

In April 1961, the Soviets blasted ahead of us by putting the first human into Earth orbit, cosmonaut Yuri Gagarin traveling in a spacecraft named Vostok 1. Being told that the Russians must be ahead of us in rocket development to achieve an Earth orbit spurred a collective sense of fright. In my college classroom, we cheered joyfully when the US finally put a man in space the following month. On May 5, Mercury astronaut, Alan Shepard, became the first American in space in a suborbital launch, prompting President Kennedy to announce a national goal to land a man on the moon before the end of the decade. America would be investing billions of dollars in the National Aeronautics and Space Administration (NASA). The Project Mercury Program and our seven Mercury astronauts gave us hope and heroes, and in February 1962, the first American, John Glenn, was launched into Earth orbit.

"We choose to go to the moon in this decade and do the other things, not because they are easy, but because they are hard. . ."

—John F. Kennedy at Rice Stadium, September 12, 1962

USS *Nautilus* SSN 571

Two years before the launch of Sputnik, an historic event took place that did not shock the world or make it on the front pages of our newspapers. But it would change the art and science of submarining—and change my personal journey.

On the morning of January 17, 1955, Commander Eugene P. Wilkinson launched *Nautilus,* the world's first nuclear-powered submarine and the sixth ship of the fleet to have that name. *Nautilus* would go on to shatter all submerged speed and distance records. Congress had authorized its construction in 1951 after the successful development of a nuclear propulsion plant by a group of scientists and engineers at the Naval Reactors Branch of the Atomic Energy Commission, led by US Navy Captain Hyman G. Rickover. President Harry S. Truman laid her keel on June 14, 1951 at the Electric Boat Shipyard in Groton, Connecticut.

Nautilus shared a name with Captain Nemo's fictional submarine in Jules Verne's classic 1870 science fiction novel, *Twenty Thousand Leagues Under the Sea,* and the USS *Nautilus,* which had served with distinction in World War II.

Nautilus catapulted the US far ahead of the Soviet Union in submarine development and warfare capabilities. While millions had their eyes fixed on space, I was captivated by her submerged achievements. President Eisenhower and the US Navy did everything they could to leverage the exploits of *Nautilus* to raise national pride and confidence in our military strength.

From July 22, 1958 to August 5, 1958, *Nautilus* crossed the Arctic Ocean from the Bering Sea to the Greenland Sea, passing submerged beneath the geographic North Pole. I eagerly read and was inspired by *Nautilus 90 North*, the narrative of the voyage published in 1959 by her commanding officer, Commander William Anderson.

Nautilus 90 North sits on my bookshelf alongside *The Silent Service, Wolfpack, Run Silent, Run Deep,* and dozens of other submarine stories.

What did I learn?

- ✓ Over 70 percent of Earth's surface is covered by oceans, seas, and bays. While the United States takes up lots of land in the Northern Hemisphere, we are an island nation surrounded by vast oceans, with the Atlantic to our east, the Pacific at our west, the Gulf of Mexico to our south, and the Arctic Ocean to the north of our state of Alaska.
- ✓ In an often-hostile world, the United States relies on sea power for our safety. (We often have to relearn that lesson.)

CHAPTER 4
QUARTERMASTER SEAMAN

Image courtesy of The Salute Uniforms (Uniforms-4u.com)

Boot camp was only two weeks long. Yet I felt enormous homesickness and loneliness for Linda, my high school sweetheart. We met in junior high, then dated and danced. When I returned home, I experienced my first, sweet homecoming; I felt so good having her in my arms again. Linda did not care for my short hair. I had gotten used to it and echoed the words of the boot camp barber: “It will grow back.”

Returning to the Reserve Center, I made good progress, advancing through the apprentice rates toward seaman (paygrade E3). A hardworking student, I scored high on Navy multiple choice tests (when in doubt, select C.) It was a matter of study, passing exams, time in service, and staying out of trouble. All of which I did well.

On my short, two-year active-duty obligation, I did not qualify for placement at a lengthy Navy technical school (A School). Consequently, I chose quartermaster (QM), a rate (specialty) that I could "strike for" and achieve without attending Quartermaster A School.

QM is one of the oldest rates in the Navy: standing watch as assistants to officers of the deck and the navigator, serving as helmsman, and performing ship control, navigation, and bridge watch duties. In the army, a quartermaster is a supply and logistics person; but in the Navy, it's a navigator's assistant.

My next set of training manuals were: *Military Requirements for Petty Officer Third and Second Class* and *Quartermaster 3 & 2*. I found studying navigation very interesting. I quickly achieved my designation, and my new insignia: a ship's helm sewn above my three seaman stripes. My new Navy title was Quartermaster Seaman Francis.

I'm movin' up!

Members of the reserve are required to perform two weeks of active duty each year, and it was time for my second "tour."

Tin Can

In the summer of 1962, I received orders to a destroyer homeported in Norfolk, Virginia. Norfolk is our largest naval base on the East Coast, and the largest naval complex in the world. It's located near the site of the Civil War battle between two steel clad ships, USS Monitor and Merrimack (CSS *Virginia*).

This tour gave me an opportunity to spend some time with my hero, Uncle Ned, Navy chief and WWII veteran. Uncle Ned once told me that early in its history, some of the residents of the former confederate state did not welcome the growing US Navy presence. Some posted signs on their lawns that read: "Sailors and dogs keep off."

Landing in Norfolk, Uncle Ned picked me up at the airport with a look of pride on his face. This time my seabag arrived with me. He drove me to the pier to meet my ship, the USS *Bearss*, a Fletcher-class destroyer launched in 1943. Destroyers are often called tin cans because their hulls were built so thin (sometimes one-inch-thick steel), and sailors claimed that they were made from tin cans. This ship was named for Brigadier General Hiram I. Bearss, a Marine who received the Congressional Medal of Honor during the Philippine-American War.

USS Bearss (DD 654)

Rolling in heavy seas.
Image courtesy of NHHC, NH 85699.

I walked up the gangway with my seabag on my shoulder, saluted the ensign (flag) on the ship's stern, saluted the watch officer, and asked permission to come aboard (the etiquette I learned at the reserve center). After presenting my orders, I was escorted by the boatswain mate of the watch (BMOW) to the crews' quarters, a large, lower compartment with many pull-down canvas bunks suspended on chains. "Boats" assigned a middle bunk and a small locker to me. Finally we went topside where I was introduced to the ship's chief quartermaster.

Thank God! I was Quartermaster Seaman Francis (proudly displaying a helm on my left sleeve) rather than just Seaman Francis. Otherwise, I would have been assigned as a deck hand. A reservist's annual two weeks is called Active Duty for Training. I was relieved that *Bearss* was observing the training part.

The young men I met in the QM gang all appeared professional and high-spirited. The chief quartermaster told me we'd be getting underway in a few days, and I would stand underway watch with them.

Two days later as we left the pier, I felt excitement because this was the first time I'd set sail on a Navy warship. I was enthralled with the navigation work being accomplished by the "gang" and their ability to keep the ship on course and in the middle of the channel. I felt good about the Navy rate I had chosen; I looked forward to someday performing as well as these young QMs.

My satisfying experience came to a sudden end as the ship began to roll side to side in the North Atlantic swells. We were steaming north to make port calls in Newport, Rhode Island, and

Boston, Massachusetts. Seasickness was gripping me, so I beat a hasty retreat to my bunk to lie horizontally, which always lessens motion sickness symptoms for me. Not yet assigned watch duty, I hoped I could ride it out in my bunk. That was not to be.

Shortly after I settled in, the BMOW appeared at my bunk, beat on it with his night stick, and ordered me to get up. I was told it's "up all bunks" during daylight hours. This was going to be difficult for me.

I staggered topside to the navigation space, wobbling back and forth with the ship's annoying motion, and reported my dilemma.

"The corpsman can issue you Dramamine," I was advised. Great!

Living on Dramamine helped me function reasonably well on watch, despite the drowsiness and the overwhelming desire to crawl back into my bunk. When we steamed into Narragansett Bay and tied at the Newport Naval Station, I was greatly relieved. Wonderful! A few days without a heaving deck under my feet.

Newport, Rhode Island, is a beautiful seaside resort city located on Narragansett Bay, noted for historic mansions, a rich sailing history, and as a major port city in the eighteenth century. Naval Station Newport is an important Navy Training Center and the home of the US Naval War College and the Naval Undersea Warfare Center.

With little spending money, and being younger than the legal age to drink alcohol, I was not able to join the crew in local bars. I spent most of my time onboard writing letters, studying my courses, walking around the base, and taking in the warm sunshine, the beautiful view of the bay, and the secure feeling of solid, unmoving ground under my feet.

When underway again, heading around Cape Cod to Boston, I had learned my lesson and loaded my system with Dramamine. The ocean was calm, and I was getting my sea legs.

Wooden Ships and Iron Men

USS *Constitution*, 1907, at Boston Navy Yard, courtesy NHHC, catalog #NH 63532.

We tied up in Boston Harbor. This was my first time in New England; I was thrilled. Still a sailor of limited means, I discovered two free things: The Boston Common and the Freedom Trail—treats for this history buff.

Boston Common (known as "the Common") is a fifty-acre public park in the center of downtown Boston, Massachusetts, dating from 1634 when a Puritan colonist purchased the land to make a public park. The Common is the oldest city park in the US and was used as a camp by the British before the American Revolutionary War, before they left for the Battle of Lexington and Concord. The park was used for public hangings until 1817, most from a large oak tree that was replaced with a gallows in 1769. Four hundred people were hanged there; cattle once grazed there; and a cemetery, the Central Burying Ground, is located on the Boylston Street side. The adjacent Boston Public Garden has a four-acre pond with ducks and swans, and Swan Boats (pontoons).

I found a map of the Freedom Trail and walked the historic streets with intense interest. The Common is the starting point for the 2.5-mile Freedom Trail (sometimes referred to as the Cradle of Liberty) with sixteen historical sites. Other historic places include: The Boston State House, Park Street Church (site of many anti-slavery meetings), Granary Burying Ground (final resting place for Samuel Adams, Benjamin Franklin's parents, and others), King's Chapel and Burying Ground (Oldest Cemetery in Boston containing numerous Puritan graves and tombs), Old South Meeting House (site of mass meetings including the meeting before the Boston Tea Party), Faneuil Hall (a marketplace and site of many speeches demanding independence from Britain), and the Old North Church (built in 1723), the oldest church and the tallest steeple in Boston.

The trail crosses a bridge to Charlestown to view the Battle of Bunker Hill Monument. The final stop was the USS *Constitution*, launched in 1797. Also known as Old Ironsides, the USS *Constitution* is a wooden-hulled, three-masted heavy frigate of the

US Navy that is the world's oldest commissioned naval vessel still afloat.

Constitution was one of six frigates constructed for the young United States Navy. She saw action during the War of 1812 against the United Kingdom, when she captured numerous merchant ships and defeated five British warships. Going below decks, on *Constitution,* I was surprised that I had to bend over and duck to avoid hitting my head on the overhead! While onboard I recalled a phrase I heard often from sailors, "There once were iron men who sailed on wooden ships."

Returning to the Common to rest, I noticed a billboard, "Ted Kennedy for US Senate." His campaign slogan was: "I can do more for Massachusetts."

There was no doubt he could. After all his brother was President of the United States.

On our return leg to Norfolk, I was more productive and diligent in my training. The motion sickness drug, my sea legs, and my motivation allowed me to spend more time upright, completing my "practical factors" on navigation skills. Being very productive, I laid out a ship's course back to Norfolk, with estimated times (ETAs) along the track, and I conducted dead reckoning and electronic navigation using LORAN (short for long-range navigation), a radio navigation system developed in the US during WWII.

To get out on deck in the fresh breeze in daylight, I practiced semaphore with flags. Under a canopy of stars, I was taught how to use a sextant and attempt to determine ship's position by celestial navigation. Getting signoffs on these skills was important for me to advance to higher rates and improve my slim pay. The young men in the QM gang were eager and generous to help me learn. This shipboard experience reinforced my desire to learn more about ship's navigation and to avoid, if possible, serving aboard a Navy surface ship that constantly sways, heaves, rocks, and rolls.

After our return to Norfolk Naval Base, I departed *Bearss,* thanking my shipmates for their patient help and signoffs on my training. Before flying back to Michigan, I was able to spend an enjoyable evening with my relatives, Aunt Thelma (my dad's sister), my cousins, Susan and Johnnie, and of course, Uncle Ned. I left Norfolk with an immerse feeling of accomplishment.

What did I learn?

- ✓ Need some help? Ask for it. Seek the help and advice of others.
- ✓ No person is a rock. No one is an island.
- ✓ Many people are willing to help others succeed.
- ✓ Walking the Freedom Trail was more interesting and educational than sitting on a bar stool.

CHAPTER 5
THE BRINK OF WAR

US Military Defense-Ready Condition Levels (DEFCON)

Readiness condition	Exercise term	Readiness	Description
DEFCON 1	COCKED PISTOL	Maximum readiness. Immediate response.	Nuclear war is imminent or has already started.
DEFCON 2	FAST PACE	Armed forces ready to deploy and engage in less than six hours	Next step to nuclear war.
DEFCON 3	ROUND HOUSE	Air Force ready to mobilize in 15 minutes.	Increase in force readiness above that required for normal readiness.
DEFCON 4	DOUBLE TAKE	Above-normal readiness.	Increased intelligence watch and strengthened security measures.
DEFCON 5	FADE OUT	Normal readiness.	Lowest state of readiness.

On the flight back home to Michigan, after two weeks away, I was filled with anticipation for many reasons. I looked forward to another homecoming with Linda, my sophomore year at college, my new job as an intern at Dow Chemical Company in Midland, Michigan, and getting my wheels back—my 1957 Ford, jet-black coupe with a V8 engine.

I caught up with Linda at the wedding of her best friend. Linda looked stunning in her bridesmaid dress. The occasion felt like a fairy tale. We agreed that we would get married that winter, during my Christmas break from school.

That fall I started my part-time internship at Dow Chemical. My responsibility was to create and evaluate new formulations for automobile antifreeze. When I left my position as a cashier at a local grocery store, I took a pay cut. But I no longer worked evenings, which would be a big help with my college studies.

The Brink

My Navy training was low key and uneventful until October 1962, when I learned about DEFCON—the Defense Readiness Conditions. That's when I and millions of Americans began to live with increased fear of a nuclear war.

On October 22, 1962, President Kennedy delivered a nationwide televised address announcing the discovery of Soviet missiles on the island of Cuba, 90 miles from our shore. The president stated, "It shall be the policy of this nation to regard any nuclear missile launched from Cuba against any nation in the Western Hemisphere as an attack by the Soviet Union on the United States, requiring a full retaliatory response upon the Soviet Union."

President Kennedy directed a quarantine of all offensive military equipment bound for Cuba. Since a blockade was considered an act of war, it was called a naval quarantine. During the president's speech, a directive went out, placing all US military forces at DEFCON 3 (Round House: Increased Force Readiness). Anxiety levels, both civilian and military, were raised as well. We were notified that the Naval Reserve (us!) might be mobilized, depending on the needs of the active units and the course of events. At the reserve center, we were provided with little information on our naval activity. Not surprisingly, I learned more from the nightly news on TV than I did through naval channels. I just didn't have a high-enough security clearance or the "need to know."

On October 26, the US Strategic Air Command (B52 bombers with thermonuclear weapons) went to DEFCON 2. (FAST PACE: Next step to nuclear war, Armed forces ready to deploy in less than six hours).

Like the rest of my fellow citizens, I hoped our civilian and military authorities would prevent nuclear war. Yet I was disappointed that I was not stationed on a ship on the quarantine picket line. Strange feeling, but I regretted not being in the action. It was life on the edge. We were engulfed in deep fear and a sense of impending doom.

Will there be another tomorrow? Is this how the world is going to end?

Later, in a college political science class, with keen interest, I wrote a research paper—a deep dive into the Cuban Missile Crisis. This was a milestone event in my young naval career, and I wanted to know more about it. We were perilously close to a nuclear war (Estimated 150 million casualties and a nuclear winter in the Northern Hemisphere).

Despite candidate Kennedy's claim of a missile gap during the 1960 presidential campaign, the US far outnumbered the Soviet Union in nuclear warheads. In addition, the US was well ahead of the Soviet Union in submarine-launched nuclear missiles.

By fall 1962, the US Navy had eight nuclear ballistic missile submarines (SSBNs) in commission, each carrying sixteen Polaris missiles. By themselves, these submarines, which could remain virtually undetected in the ocean, gave the US nuclear superiority.

Why in the world would the Soviet Union place nuclear missiles in Cuba?

Blame it on the disastrous Bay of Pigs invasion in 1961. Kennedy's predecessor, President Dwight Eisenhower, had funded the US Central Intelligence Agency (CIA) to plan and execute an invasion of Cuba.

Kennedy would forever regret his decision to listen to the "hawks" on his Joint Chiefs of Staff and to authorize the invasion by Cuban exiles who were quickly defeated by the army of Fidel Castro, Cuba's communist dictator. The failed invasion created continued fear in Cuba and in the Soviet Union (Cuba's international benefactor) that the US would attempt a future invasion.

Consequently, Soviet First Secretary Nikita Khrushchev agreed to Cuba's request to place nuclear missiles on the island to deter a future invasion. The presence of Soviet missiles based in Cuba was confirmed on October 14, 1962, by photographs taken by an American U-2 spy plane.

Much has been written about the intense and adroit diplomatic gymnastics that prevented a shooting war. Kennedy assembled a group known as the Executive Committee to manage the crisis. They tirelessly worked diplomatic channels; partnered with our NATO allies, the Organization of American States, and the United Nations; and made direct contact with Soviet leaders.

In addition to his political advisers, Kennedy was surrounded by military members of the Joint Chiefs of Staff. Many "hawks" argued for military air strikes and an invasion of the island. Kennedy resisted the urging of several military advisers to bomb and invade Cuba. The president was reluctant to kill Russian soldiers or technical advisors stationed on the island.

The Naval Blockade

Prior to the Cuban Missile Crisis, US military leaders had planned for a bombing and invasion of Cuba, but they had to scurry to create a plan for a Naval Blockade. It fell to the commander in chief, Atlantic Fleet, to deploy ships and other forces to enforce the quarantine.

The US Navy's mission was to positively locate ships enroute to Cuba and prepare for interception when they crossed the quarantine line. The first intercept was the *Kimovsk,* suspected of carrying military equipment, and the second was to be the *Poltova* with a possible cargo of missiles. During daylight hours and when in visual signal distance, a US Navy warship was to be dispatched to a position close aboard, but which would not give the Soviet ship an opportunity to ram. Our ship would then display by visual flag hoist the international signal "K" (stop at once) or "ON" (heave to at once). These signals were to be paralleled by signal light.

If this failed to halt the intercepted ship, minimum force was authorized to damage non-vital parts of the ship, but to refrain if possible from personnel injury or loss of life.

Once the ship was stopped, a party, including Russian linguists, were to board the ship. "Visit and search" was to include examination of the manifest and inspection of the cargo. In the event visit was refused, the ship was to be taken into custody and forcefully boarded to control the ship's operation.

My cousin Kenny, grandson of Arnold and also a Michigan naval reservist, went on active duty a few years ahead of me. Kenny had served as a torpedoman aboard the Forest-Sherman class destroyer USS *Barry*, but left the ship just before *Barry* was deployed to the picket line. On October 16, 1962, the day President John F. Kennedy was shown aerial reconnaissance photographs of Soviet nuclear missiles and launch sites under construction in Cuba, *Barry* was still undergoing upkeep at Newport.

On 22 October, when President Kennedy told the nation that he had initiated a "strict quarantine of all offensive military equipment under shipment to Cuba," *Barry* cleared Newport that night in company with *Blandy*, *Charles S. Sperry*, and *Keppler*. After rendezvous with US Aircraft Carrier *Essex* on the 26th, she operated as a screening vessel and plane guard.

Two days later, *Barry* was detached to operate on anti-submarine warfare surveillance and kept a close watch on contact C-19, a surfaced Soviet submarine. *Barry,* well east of the quarantine line, kept the foxtrot-class diesel boat under surveillance until it submerged that evening.

Barry remained on the line, carrying out patrols until November 8, when she was ordered to investigate a soviet merchantman. Sighting the merchant ship the next evening. *Barry* closed to within 400 yards on the merchantman's starboard quarter, illuminated the ships' quarter and bow, and identified her as the Soviet-registry *Metallurg Anosov*. Trailing astern, *Barry* followed the merchant ship, heading east away from the quarantine zone, until morning. After dawn, the destroyer closed the merchant to take photos of deck cargo before she shaped course for *Essex* for refueling.

A Close Call

Submariners were the ones who caused a situation that brought us closest to the brink of war. On October 1, 1962, four Russian Foxtrot diesel electric submarines deployed from the Kola Peninsula northwest of Murmansk headed for Mariel port in Cuba to establish a submarine base. Each submarine carried nuclear-tipped torpedoes (one was Soviet Submarine B-59).

On October 27, units of the US Navy aircraft carrier USS *Randolph* and eleven destroyers detected B-59 near Cuba. US vessels began dropping depth charges of the type used for naval training, containing very little charge, not intended to cause damage but to communicate with the submarine. The purpose was to force it to surface for positive identification.

But messages from the US Navy to communicate that practice depth charges were being used never reached B-59 or Soviet Naval Headquarters. B-59 had not been in contact with Moscow for a number of days; and although the submarine's crew had earlier picked up US civilian radio broadcasts, once they began attempting to hide from pursuers, the vessel had to run too deep to monitor any radio traffic. So those on board did not know whether or not war had broken out.

Believing that war had started, the captain of the submarine, Valentin Grigorievich Savitsky, wanted to launch a nuclear torpedo. But B-59 was the only submarine in the flotilla authorized to launch the torpedo if all three senior officers on board unanimously agreed to do so.

Commander of the deployed submarine detachment, Vasily Arkhipov, who was equal in rank to Savitsky, opposed the launch and eventually persuaded Savitsky to surface and await orders from Moscow. Once the submarine's batteries ran low and its air-conditioning had failed, B-59 had to surface amid the US warships pursuing it; then it set course back to the Soviet Union.

Along with all humanity, I genuinely feared atomic death. In late November 1962, the confrontation was resolved with an agreement with the Soviet Union—they would remove their missiles from Cuba in exchange for the removal of US medium-range missiles in Turkey and Italy. Nuclear war didn't happen, and we were free to get on with our lives.

After learning about the end of the crisis on the evening news, Linda and I fell into a deep and thankful embrace. We would survive and make it to our wedding day.

The Cuban Missile Crisis renewed my resolve to help win this Cold War.

What did I learn?

- ✓ If the political hawks had had their way in 1962, the US would have gone to war. Americans were fortunate that John Kennedy was a critical thinker and found a way to resolve the conflict without dropping bombs, launching missiles, or invading Cuba.
- ✓ "Events of October 1962 indicated, as they had all through history, that control of the sea means security. Control of the seas can mean peace. Control of the seas can mean victory. The United States must control the seas if it is to protect your security . . ." —President John Kennedy.
- ✓ The world is often hostile and dangerous. It helps to hold onto hope and hug our loved ones.

CHAPTER 6
WILL YOU TAKE THIS SAILOR?

In the summer of 1962, Linda and I, two American kids from the heartland, had agreed to marry on my winter break from college. We were, after all, employed, in love, and semi-independent adults. In many ways, the Cuban Missile Crisis felt like a near-death experience. Having avoided nuclear war confirmed our decision to marry. And don't forget, in the 1960s, it wasn't considered acceptable to just move in together.

Following an ancient tradition of Irish men, I married up. The evening of December 29 was beautiful and cold. The sky was clear, and the snow glistened with starlight. Linda and her family arranged the wedding ceremony at their Episcopal Church on Bay City's west side and planned a fine reception at their home. My main responsibility was to become an Episcopalian.

When we vowed to get married, Linda's only request of me was that I become an Episcopalian. I accepted and agreed to work hard and learn how to spell E-P-I-S-C-O-P-A-L-I-A-N. Raised in the Baptist church, I needed to complete a confirmation class with the Episcopal priest. I did, but wondered why the *Episcopal Book of Common Prayer* was still written in sixteenth century Elizabethan English.

Linda and I were surrounded and loved by many relatives and friends, including my Uncle Ned wearing his Navy dress blues as we heard, "Dearly beloved: We have come together in the presence of

God to witness and bless the joining together of this man and this woman in holy matrimony."

We nineteen-year-old newlyweds moved into a second-floor apartment where we struggled to keep it all together on a slim budget and busy schedules. With Linda's job at the library and mine at Dow Chemical, our combined income was $62 dollars per week. Fortunately, Linda was a wizard at budgeting. That was the start of many years when she kept us afloat on my meager paychecks. We bought three rooms of furniture for about $600. Linda decorated our first apartment beautifully and placed in our bare cupboards the few donated and purchased dishes and utensils we possessed. Since Linda frequently worked evenings at the library, many days we were like ships passing in the night, but we were together!

Linda became my partner and my homework helper and savior. She typed much faster than I could. Often late at night, because I procrastinated, she typed my assignments and papers on our manual Royal typewriter.

"I'll put the coffee on, honey."

In one research paper, I wrote about the active civil rights movement in the American South. It was 1963 during the height of the battle for equal rights for all Americans. I chose not to write about the most prominent civil rights leaders, often called the Big Six: labor organizer Asa Philip Randolph; Dr. Martin Luther King, Jr. of the Southern Christian Leadership Conference; James Farmer Jr. of the Congress Of Racial Equality; John Lewis of the Student Nonviolent Coordinating Committee (SNCC); the National Urban

League's Whitney Young, Jr.; and Roy Wilkins of the National Association for the Advancement of Colored People (The NAACP.)[5]

Instead I chose to write about Medgar Evers, probably because Medgar was a veteran of World War II. He was killed, not on the battlefields of Europe, but in front of his house in Jackson, Mississippi. He was killed fighting for freedom, not by a German bullet in France, but by an American bullet in Mississippi. The title of my paper was, "Murder in Mississippi."

Evers' life story demonstrated how difficult and dangerous it was for a Black person to live in the American South and work for civil rights in the violent 1960s. Medgar Evers was born on July 2, 1925, in Decatur, Mississippi, the third of five children. The Evers family owned a small farm. Medgar worked on the farm and at a sawmill. He and his siblings walked 12 miles a day to attend segregated schools. Eventually Medgar earned his high school diploma. He served in the US Army during World War II from 1943 to 1945. He was sent to the European Theater where he fought in the Battle of Normandy in June 1944. After the end of the war, he was honorably discharged as a sergeant.

In 1948, Evers enrolled at Alcorn Agricultural and Mechanical College (an historically Black college, now Alcorn State University), majoring in business administration. He earned his Bachelor of Arts in 1952. In 1954 he was named as the NAACP's first field secretary for Mississippi. He organized boycotts, established new local

---

[5] The Big Six organized the March on Washington for Jobs and Freedom, held August 1963, where Dr. King gave his famous, "I have a dream" speech on the steps of the Lincoln Memorial.

chapters of the NAACP, and worked for long-denied negro voting rights.

His civil rights leadership in Mississippi made him a target of White supremacists. His public investigations into the 1955 lynching of Chicago teenager Emmett Till in Mississippi made Medgar Evers a prominent Black leader who lived with a constant threat of death.

On May 28, 1963, a Molotov cocktail was thrown into the carport of his home. On June 7, 1963, Evers was nearly run down by a car after he came out of the NAACP office in Jackson, Mississippi. A large White supremacist population and the Ku Klux Klan were present in Jackson and its suburbs. The risk was so high, Evers and his wife Myrlie had trained their children on what to do in case of a shooting, bombing, or other kind of attack on their lives.

On Wednesday, June 12, 1963, just hours after President John F. Kennedy's nationally televised Civil Rights address, Evers pulled into his driveway after returning from a meeting with NAACP lawyers. Evers' family was waiting for him, and his children exclaimed to his wife Myrlie that he had arrived. Emerging from his car and carrying NAACP T-shirts that read "Jim Crow Must Go," Evers was struck in the back with a bullet.

Myrlie was the first to find him. He was taken to the local hospital in Jackson, where he was initially refused entry because of his race. His family explained who he was, and he was admitted and died in the hospital fifty minutes later. He was only thirty-seven years old.

Medgar Evers was the first African American to be admitted to an all-White hospital in Mississippi. Mourned nationally, he was buried on June 19 in Arlington National Cemetery, where he received full military honors.

People were dying because they had the courage and the will to work for their civil rights and racial justice. In 1963, I felt an urge to join the freedom marches and demonstrations—to be relevant, to make a difference, to help "bend the long arc of the moral universe toward justice." But I chose a related but different path by joining the US Navy. I consoled myself and rationalized that I played a role in defending our freedom, however unequal it was. Linda and I had no clue how much more tragedy 1963 would bring.

The USS *Thresher* Tragedy

USS *Thresher* (SSN 593) Painting, courtesy of United States Naval Institute.

The USS *Thresher* SSN 593 was the lead ship of a new class of US Navy nuclear fast-attack submarines. Created to find and destroy Soviet submarines, *Thresher* was the fastest and quietest submarine of its day. She also had the most advanced weapons system, including launchers for the US Navy's newest anti-submarine missile, the SUBROC, as well as passive and active sonar that could detect vessels at unprecedented range.

Thresher got underway with 129 crew and shipyard personnel aboard for her "shakedown" post-availability sea trials from Portsmouth Naval Shipyard on April 9, 1963, accompanied by the submarine rescue ship *Skylark*. During her trial, *Thresher* remained in contact with *Skylark* and reported her progress as she incrementally dove to test depth, pause, and check her systems' integrity. Reaching her test depth at 9:13 a.m., *Skylark* heard garbled communications from *Thresher* via the underwater hydrophone, ". . . experiencing minor difficulty, have positive up-angle, attempting to blow . . ." *Skylark* received another communication at 9:15 a.m., this time completely unintelligible.

Skylark queried *Thresher*, "Are you in control?"

With no response. At 9:17 a.m. *Skylark* received her final communication from *Thresher,* also garbled, but ". . . test depth . . ." was heard. *Skylark* personnel subsequently detected sounds in the water indicative of an implosion.

One *Skylark* operator stated, "I had heard a lot of ships breaking up during World War II after having been torpedoed . . ."

Efforts to reestablish contact with *Thresher* failed, and a search group was formed to locate the submarine, unsuccessfully. Rescue and salvage ship *Recovery* subsequently recovered bits of debris, including gloves and internal insulation pieces. Photographs later taken proved that the submarine had broken up, taking all 112 crewmembers, and seventeen civilian personnel aboard to their deaths in 8,400 feet of water, 220 miles east of the Massachusetts coast.

Driving home from community college, I learned about the loss of the *Thresher* on my car radio. Together, Linda and I watched the many stories on the evening TV news. We didn't talk much about it. I confessed that submarine service is risky. "That's why submariners earn hazardous duty pay."

Feeling young and indestructible, I thought, *It won't happen to me.*

The following year, while in enlisted submarine school, I would learn more about what caused the catastrophic loss and tragedy as well as the Navy's efforts to make US submarines safer.

November 22–25, 1963

On Friday afternoon, November 22, Linda was putting books on the shelf at her library when the elderly head librarian rushed to her to say, "Roosevelt has been shot." She meant, of course, President Kennedy had been shot. The slip of the tongue was telling. My generation was devastated by the death of JFK, just as hers was at the death of FDR.

The tragic news came to me in my laboratory at Dow. A hero of mine was just shot dead. I felt shock and despair. That weekend was agony. Saturday was my best friend's wedding, and Sunday was my parents' twenty-fifth wedding anniversary party. We did not want to party; we wanted to be one with the television and mourn. Linda and I tried to console each other and lamely attended the required events.

When we were able to "tune in," we saw the relentless television images. President Kennedy was shot by an assassin in Dallas, Texas. First Lady Jacqueline Kennedy, riding with him in an open convertible, terrified. The killer, Lee Harvey Oswald, later arrested, then shot and killed, while in police custody, by Jack Ruby—all shown on TV.

After Vice President Lyndon Johnson was sworn in as the 36th President of the United States aboard Air Force One, and JFK's body was flown back to Washington DC, Jackie Kennedy planned a majestic funeral and burial for her husband that gave the nation and her an opportunity to honor and grieve. This indelible and consequential experience, November 22, 1963, was, for me and my generation, our Pearl Harbor Day.

Prior to my two-year active-duty time, I requested orders to US Naval Enlisted Submarine School. Informed that I needed to extend my enlistment by six weeks, I quickly agreed. Working part-time, I earned my two-year degree in three years. Snow was still on the ground in Michigan when I received my associate degree in April 1964.

Seaman Francis, reporting for active duty in May, could not afford to support a wife until he advanced to a higher pay grade. Consequently, Linda and I were separated for several months. We moved our three rooms of furniture into a spare bedroom in her parent's house. Linda redecorated her former bedroom and moved in with her parents.

What did I learn?

- ✓ Love, compassion, and loyalty are powerful human qualities.
- ✓ Never miss an opportunity to hug your loved ones.
- ✓ Having a cause greater than self-interest can create a life of purpose.
- ✓ It's not what we gather, but what we scatter that will define our lives.

CHAPTER 7
GROTON ON THE THAMES

Enlisted Submarine School Class
(Seaman Francis, second row, second from right.)

The Naval Submarine Base New London is located on the Thames River in Groton, Connecticut, across from New London, along the Long Island Sound. To be a submariner, you need to pass through the Naval Submarine Base New London.

As requested, I received orders to the US Naval Enlisted Submarine School along with my customary travel vouchers. Linda drove me to the airport. We shared a painful goodbye. Leaving her, I felt a sense of foreboding and loneliness. This emotion would visit me again and again as our separations became longer and more frequent. Off I went in my crisp white uniform.

My flight landed at LaGuardia Airport in New York, and I was pleased that my seabag also landed at LaGuardia as well. Being a novice traveler, I walked out the front door of the airport terminal and approached the cab nearest me for a ride to Grand Central Station. Abruptly an officer redirected me to go to the front of the cab line. *Oh, so that's the rule.*

The grand size of Grand Central Station made my eyes widen. Perusing my travel orders, I located the boarding ramp for the New York, New Haven, and Hartford Railroad train to New London. I had a few hours to spare when a random piece of information sprouted into my head. The legal drinking age in New York State was eighteen; I was now twenty years old, and a sailor.

On the window of the nearest bar was a sign for Schaefer beer. And I even thought of their jingle: "*Schaefer is the one beer to have when you're having more than one.*" As I settled on a vacant stool at the bar, the man next to me greeted me and asked where I was stationed.

"I'm on my way to Submarine School in New London."

"Can I buy you a drink?"

I thanked him and suggested a Schaefer beer.

After I finished my free beer, I said thank you and goodbye to my new friend. As I stepped onto the floor, he said, "Did you fall into a hole?" *Oh my god. Another "short" joke.*

On the train trip north to New London, I sat on a hard wooden seat and looked outside at the passing scenery. The Connecticut structures we passed were older than those I was accustomed to seeing in Michigan. Not surprising, since New England is much older than the Midwest. After I arrived at the railroad station in New London, I remembered the protocol and walked to the head of the cab line.

The cab let me off at the submarine base entrance. I presented my Navy ID Card and orders to the Marine guard at security. He directed me to the enlisted barracks. I carried my heavy seabag on my shoulder while looking at the base map for directions.

After I checked into the barracks, I confirmed with the barrack's watch the location of my classroom. The barracks was a pure World War II-vintage, wooden structure. After breakfast the following morning at the Enlisted Mess Hall, I arrived early at school. Nearly all of my enlisted Submarine School classmates wore Navy seaman stripes, like me.

Being spring-loaded from my bootcamp experience, I expected some planned military abuse. But there was none, and I was relieved that everyone was respectful and treated me like I actually had a brain. For many weeks, I relished the classroom learning about submarine construction and operations.

A number of my classmates were Pakistani sailors. The US was in the process of turning over one of our older diesel-electric submarines to their navy. Their English was excellent, and we had many enjoyable conversations. I asked one Pakistani sailor what they planned to do with their new submarine.

He said, "India has an aircraft carrier. We will sink it."

I thought, *Indeed, we are the arms merchant to the world.*

The escape tower looms high above all the other buildings on the submarine base. I looked at it often and dreaded the day I would be required to enter the tower to make a free ascent. A long submarine tradition, the free ascent is designed to train submarine sailors for a scary escape from a sunken submarine.

To prepare us for this ordeal, we underwent pressure testing. First we stripped down to our underwear and sat in a chamber. The instructor closed the door then increased air pressure gradually, giving us time to equalize the pressure to our sensitive ear drums. The air temperature became warm and then hot. Then the operator quickly released the pressure.

I worked desperately to equalize and cope with the falling air pressure and the dropping temperature. Ice formed on the small windows. We all passed our pressure test. Next was the free ascent in the escape tower.

Escape Tower,
Submarine Base, New London

The dreaded day dawned, and we marched to the tower. We received a classroom briefing by a group of fit Navy divers. Several times they repeated the warning that in a free ascent, it was critical that we blow all of the air out of our lungs.

They showed us on a graphic how excess nitrogen bubbles form in our bodies if we do not adequately exhale as we quickly rise to the surface. If the transition occurred too fast, nitrogen doesn't have the time to clear from your blood, and it forms bubbles in the tissue or blood that cause decompression sickness, the bends. The ensuing joint and bone pains can be severe enough to double you over. We were told that would be painful and life-threatening.

Following our lecture, we were led to an elevator and proceeded down to the 50-foot level. There is also a 100-foot level, and I was relieved that we didn't go that low. My small group was crammed into a pressure chamber attached to the outside of the water tower. And our instructor began filling the chamber with water. I plugged my nose and equalized to the increasing pressure in the compressed atmosphere. The water reached my chin, so I grabbed a valve and pulled my body up.

I followed the firm instructions on the escape procedure: "Duck down. Step out of the chamber onto the platform and inflate your life preserver. Hold your hands above your head and begin a strong blow. You will not be released until you have a strong blow!"

I was on my way to the surface as though riding a fast elevator.

Before I reached the surface, I felt empty, as if I had no more air left in my lungs, and I made a bad decision and stopped blowing. Immediately I was grabbed by a Navy scuba diver who poked his hand forcefully into my stomach, and I began to release air again. I guess I did have some air left.

The diver rode with me to the surface and pulled me aside. "What's your name sailor?"

"Francis," I replied.

"Francis. You will need to do this twice to pass."

It's true. Practice makes perfect. My next two ascents went as required.

In addition to simulated escape training from a sunken submarine, we learned about damage control and firefighting. All shipboard fires are dangerous. But a fire on a submerged submarine is potentially catastrophic because of the sealed environment. Different methods are required.

For example, you don't put out an electrical fire with water. You use carbon dioxide, foams, or dry chemicals. Wearing an uncomfortable gas mask, I got to battle several fires with several agents in the enclosed bunker. It was noisy, chaotic, and hard to see in front of me.

To keep us humble, we were required to remove our gas masks in a building filled with smoke, and another time when the building was filled with tear gas. Naturally I choked and suffered from burning and watering eyes. It drove home to me the danger posed by fire aboard a submarine.

USS *Clamagore* SS 343

USS Clamagore off the Philadelphia Naval Shipyard, courtesy NHHC, catalog #NH 85093.

For much of US Navy submarine history, our boats were named after fish and other sea creatures (*Perch, Trout, Shark, Thresher, Seawolf,* etc.). That changed with different missions and politics (ballistic missile submarines: *George Washington, Thomas Edison, Sam Rayburn,* and many others). We even named a few submarines after confederate generals (*Robert E. Lee* and *Stonewall Jackson*).

The clamagore is a blue parrot fish found in the Caribbean, and the name of the first submarine I went to sea on. *Clamagore* was built by the Electric Boat Company in Groton, Connecticut, near the end of World War II. She was launched on February 25, 1945.

In 1962, *Clamagore* became one of nine boats to undergo the Greater Underwater Propulsion Power Program (GUPPY) conversion. World War II-era fleet submarines underwent conversion to improve submerged speed, maneuverability, and endurance. *Clamagore* finished her GUPPY III conversion in February 1963 and was transferred to Submarine Squadron 2 in Groton, Connecticut.

The Navy began the GUPPY program by testing and reverse-engineering two captured German Type XXI U-boats. That analysis led to four goals—increasing the submarines' battery capacity, streamlining the boats' structures, adding snorkels, and improving fire control systems. The snorkel allows a submarine to draw outside air into the boat to run the diesel engines while submerged at periscope depth. It was developed by Dutch engineers and widely used on German U-boats during the last year of WWII and known to them as the *Schnorchel.*

In summer 1964, our enlisted submarine school class got underway from Groton on the USS *Clamagore* and headed out into Block Island Sound. I felt like a kid at a candy store as I heard the diving alarm, live, for the first time. As she took a down angle and submerged, I braced myself. Each of us was allowed up into the conning tower to look through the periscope.

She was cramped and smelled of diesel fumes similar to the submarine I toured back in Bay City. I walked fore and aft, examining the equipment and operations of the boat, and speaking with crew members. The seas were calm in Block Island Sound that day, so I experienced no motion sickness. But I did feel the exhilaration of being at sea on my first submarine. Returning to Groton, I was even more anxious for graduation and assignment to my first boat.

SUBSAFE

In class one day, the instructor discussed the tragic loss of the USS *Thresher*. *Thresher* lies in six major sections at 8,400 feet on the ocean floor. The major sections are the sail, sonar dome, bow section, engineering spaces, operations spaces, and the tail section.

As you recall, *Thresher* got underway with 129 crew and shipyard personnel aboard for her "shakedown" post-availability sea trials from Portsmouth Naval Shipyard on April 9, 1963.

While making a test depth dive to 1,300 feet, flooding occurred in the engine compartment. Nuclear reactor procedures at the time required that the reactor operators shut down (scram) the reactor and closed the main steam valve. *Thresher* was then deprived of steam propulsion.

She commenced main ballast tank blow and was reducing depth. However, the ballast tank blow reducers, which reduce high pressure air to low pressure, began to freeze up in the cold water. Ballast tank blow was interrupted. As water rushed into the engine compartment, *Thresher* became negatively buoyant and began to slide back into the depths. She was crushed by enormous sea pressure.

The US Navy's post-loss inquiry uncovered design, construction, quality assurance, and operations flaws. Numerous remedial actions were initiated including the submarine safety (SUBSAFE) program to improve submarine safety and survivability.[6]

---

[6] Navy releases documents from 1963 USS Thresher sinking, deadliest submarine disaster in US history, by David Sharp, The Associated Press Sept 24, 2020, in *Navy Times*

I was relieved to learn about the SUBSAFE program.

I first began to experience the "band of brothers" feeling at Enlisted Submarine School during the eight-week training with sailors from many other parts of the United States. Some were raised in the South—unfamiliar territory to me. I played chess with and enjoyed the company of James, a classmate from North Carolina. He was a graduate of Wake Forest University, polite and well-mannered. He reminded me of Ashley Wilkes, from Margaret Mitchell's 1936 novel, *Gone with the Wind*,

At graduation, I wished I could join my classmates, like James, who immediately received orders to an active submarine. Though I finished in the top 10 percent of my class, as a reservist, I was sent back into the Naval Reserve system to await further assignment. After a short week's leave back home, I received orders to the Philadelphia Naval Receiving Station.

Philadelphia, PA

I admit I had a bad attitude about my assignment to Philadelphia. The weather was hot, humidity oppressive, and no air conditioning in the old barracks. I felt like I was wasting my time and the Navy's money. When I reported to the naval station, I was not assigned any duty except to wait for further orders. The benefit of that later occurred to me. I could use this "dead" time to work on my self-

---

https://www.navytimes.com/news/your-navy/2020/09/24/navy-releases-documents-from-1963-uss-thresher-sinking-deadliest-submarine-disaster-in-us-history study for advancement to quartermaster third class (QM-3) and also visit the historic Colonial American sites in downtown Philadelphia.

Doing both, I visited the Liberty Bell and Independence Hall, completed in 1753 as the Pennsylvania State House where both the Declaration of Independence and the US Constitution were debated and adopted. Our courageous founders, who wore wigs and funny pants, established a new nation and "lit the torch of freedom for nations then unborn."

Being in the place where our country's liberty was first proclaimed was humbling and inspiring. I finally received orders to my first submarine and left Philly just before the 1964 race riot in North Philadelphia over a three-day weekend in late August.[7]

What did I learn?

- ✓ Look for opportunities in adversity and in diversity.
- ✓ Freedom is not free.
- ✓ The American Experiment, our fragile democracy, needs constant care and feeding if we are to keep it.

[7] At that time in Philadelphia, 18% of the city's population were Black, police brutality was rampant, and 40% of the city's jails and prison inmates were Black. Philadelphia Race Riot (1964) by Suyent Rodriguez Candeaux, *Black Past,* Dec. 7, 2017. https://www.blackpast.org/african-american-history/philadelphia-race-riot-1964/

CHAPTER 8
Seawolf

USS *Seawolf* (SSN-575) in the Thames River, New London, Connecticut, 1967, courtesy NHHC, catalog #USN 1130265, and *Seawolf* ship patch Insignia, courtesy NHHC, #NH 65602-KN.

Seawolf is solitary fish with strong, prominent teeth and projecting tusks that give it a savage look.

USS *Seawolf* General Characteristics

- 3,260 tons surfaced, 4,150 tons submerged
- 350 feet long
- Speed: 23 knots surfaced, 19 knots submerged
- 101 officers and men
- Six 21-inch torpedo tubes forward

My orders were to the USS *Seawolf*. *Seawolf* was the US Navy's second nuclear-powered submarine, built after the *Nautilus*. She was commissioned in 1957 as a fast-attack sub. But by the time I arrived on board seven years later (1964), she was a slow, noisy, target submarine. *Seawolf* was famous in the fleet for several reasons:

1) She was built with an experimental nuclear reactor, a liquid metal-cooled (sodium) nuclear reactor known as the Submarine Liquid Metal Fast Reactor that was designed to provide greater power than the water-cooled S2W reactor in the *Nautilus*. However, with too many operational, and maintenance problems, the reactor was later replaced with a pressurized water reactor like in the *Nautilus*.
2) Following successful sea trials in 1957, *Seawolf* steamed over 6,000 nautical miles to participate in NATO exercises. She surfaced off Newport, Rhode Island, and picked up President

Eisenhower, making him the first US president to go to sea in a nuclear submarine.

3) Back in 1953, Lieutenant James Earl "Jimmy" Carter, Jr. (USNA class of 1946) prepared to become the engineering officer on *Seawolf*. Carter was a Navy nuclear power-trained officer. However, when his father died in July 1953, Carter resigned from the Navy and returned to Georgia to manage his family farm. At this writing, Jimmy Carter, our 39th president, is the only submarine officer to be elected President of the United States. The USS *Jimmy Carter*, named in his honor, is a *Seawolf* class submarine, the lead ship of the US Navy's latest fast-attack family. Jimmy Carter inspired me in the same way that John Kennedy did. Carter devoted his life to public service. He consistently appealed to our "better angels." And whaddya know? They were both US Navy men!

**Lieutenant James Earl (Jimmy) Carter Jr.
in USS K-1 (SSK-1) main control room, courtesy NHHC**

4) In 1963, *Seawolf* participated in the search for the lost USS *Thresher* and in various local and fleet operations until April 1964. On 28 April, *Seawolf* stood out of New London en route to the Mediterranean Sea and a 3.5-month deployment with the Sixth Fleet. During that period, *Seawolf* operated with the aircraft carrier *Enterprise*, guided-missile cruiser *Long Beach*, and guided-missile frigate *Bainbridge,* as a part of the world's first nuclear task force.

Just after *Seawolf* returned from her Mediterranean deployment, I reported abroad. She was moored at State Pier on the New London side of the Thames River, outboard of the submarine tender USS *Fulton*. I presented my orders to her executive officer and was introduced to the chief of the boat (COB). The COB is the submarine services' version of a general foreman who is generally the senior enlisted person aboard responsible for enlisted crew performance.

I received a warm welcome and felt relieved when I was assigned to the quartermaster gang. *Seawolf's* quartermaster chief would become a mentor and take a genuine interest in my training and advancement. Then the bad news: there were more sailors than bunks on *Seawolf,* so I was assigned to the torpedo room and shown how to make a makeshift bunk on the torpedo skids that support the torpedoes. Appropriately named a skid bunk.

With no assigned locker, I had to live out of my seabag and move my canvas rack around to stay clear of frequent torpedo room operations. With some nostalgia, I actually thought back to the destroyer I rode the year before. At least I had a stationary rack then. But I was not alone. There were two other seamen with the same primitive accommodations.

Later I met the petty officers on the QM team. They all showed a willingness to take me in and train me. *Seawolf* was spacious compared to the old diesel boats. The air was pure (no diesel fuel fumes), but the cool, metallic smell was not like fresh air. As expected, I was given my submarine qualification notebook. This contained the knowledge requirements I needed to master in order to earn my submarine "silver dolphins," the warfare chest pin worn proudly by enlisted submariners to show that they know their stuff and are qualified to be a member of a submarine crew.

"Life is simple—either you're qualified or you're not!"

Dolphins - Submarine Warfare Insignia, photo used with permission, Gary Flynn, CWO4 (SS) USN (Retired)

History of the Submariner's Dolphins

The insignia of the US submarine service is a submarine flanked by two dolphins. Dolphins, attendants to *Poseidon*, Greek god of the sea, and patron deity to sailors have a characteristic way they dive and surface.

On 13 June 1923, Captain E. J. King, commander, Submarine Division Three (later fleet admiral and commander in chief, US Fleet, during WWII), suggested to the Secretary of the Navy that a distinguishing device for qualified submariners be adopted. He submitted a pen-and-ink sketch of his own, showing a shield mounted on the beam ends of a submarine, with dolphins forward of, and abaft, the conning tower.

The Philadelphia firm, Bailey, Banks & Biddle, which had done work in the field of Naval crests, was approached by the Bureau of Navigation to design a suitable badge. Two designs were submitted by the firm, one of which was recommended to the Secretary of the Navy in March 1924 by the chief of the Bureau of Navigation. The recommendation was accepted by Theodore Roosevelt, Jr., acting Secretary of the Navy.

The Submarine Crew Food Chain

Submarine captain
- Executive officer
  - Ship's officers
    - Chiefs
      - Senior enlisted
        - Qualified petty officers
          - Unqualified petty officers
            - Qualified seamen
              - Unqualified seamen (me)

Soon it became obvious to me that my existence would be at the very bottom of the silent service hierarchy, an unqualified seaman with no permanent bed to crawl into. I assessed my new circumstances and was mentally distressed. In the US Navy food chain I was lower than whale droppings. And I *volunteered* for this!

My wife Linda was back home in Michigan because I could not afford to support her. Lonely and sad, I recalled what I overheard a crusty, old Boatswain mate on the Destroyer Bearss say, "You were not issued a wife in your seabag."

This was also the US Navy in the early 1960s, before all-volunteer military service. It was commonly accepted that a sailor could be disciplined, disparaged, and outright yelled at in the presence of others, which would happen to me a few times. But a least the US Navy had discontinued floggings by cat-o-nine tails (abolished in 1850).

But the reality that I could be upwardly mobile in this Navy caste system gave me hope and motivation. I believed in progress and possibilities. To move up, I needed to 1) earn my dolphins by completing my submarine qualification program, and 2) learn the skills of my "new trade," pass the exam, and advance to Navy petty officer. There was hope that I could earn more respect and more money.

As an experienced and disciplined student, I knew the drill: Ask questions, read, and learn. Study hard, work hard, follow the rules, and seek help. I would harness my considerable persistence, insane focus, and native intelligence; go to work and climb the ladders.

The Notebooks

At my side nearly at all times, I carried two notebooks and a piping manual. Piping manuals, produced by the ship builders, contain diagrams of the ship's systems (electrical, water, hydraulics, etc.). *Seawolf*'s piping manual was made by Electric Boat, Groton. I studied it, traced the actual systems on board, and used it to answer numerous qualification questions.

The submarine qualification notebook contained certification sheets for everything I needed to learn about the USS *Seawolf.* The other notebook was for my practical factors: the sign-off cards for the knowledge and skills I needed to master in the quartermaster rate. Some practical factors were fairly easy, like plotting a ship's position. Others I found more challenging such as Morse Code, International Maritime Signal Flags (nautical flags), semaphore, and flashing light.

For many hours I practiced semaphore with Petty Officer Brown, "Browny," as he was known. *Panama Banana* was a popular drill. Browny was actually a signalman qualified as a navigator's assistant. He became my "sea daddy" and took an active role in my training. He was a sailor's sailor, highly competent and widely respected. And fortunately he was a patient teacher.

My first time going to sea on *Seawolf,* as I climbed the cold, steel, vertical ladder and reached the bridge, I could see the Thames River covered in fog. *Seawolf*'s chief of the boat assigned me as the starboard lookout. My first underway on a nuclear submarine, and I was filled with a sense of adventure.

Underway was called the maneuvering watch. The maneuvering watch officer of the deck (OOD) was a naval academy graduate and a nuclear-power trained officer, the cream of the Navy, but he was not as arrogant as some of the others. In curiosity, I asked him how we would navigate safely through the fog. He treated me respectfully as a young student rather than a lowly seaman. He said we would navigate by radar and visually. Lookouts like me were important to ships' safety. And he thought the fog would burn off by the time we got into the sound.

Ship's position reports would come from the chief quartermaster and his team stationed in the conning tower. *Seawolf* had a conning tower where the periscope was located similar to our older diesel electric submarines. Two tugboats came alongside to pull us out into the center of the river channel.

The fog was still thick as we headed out to sea. I scanned the starboard side with my binoculars and naked eyes from bow to stern, but I could not see very far and saw no other ships or boats. We steamed out of the river and into the sound, and I could feel some slight ship's motion. Radar reported no contacts close aboard. The OOD ordered rig for dive. When the chief of the watch (Yes, the COW) reported the ship rigged for dive, we were ordered off the bridge. After I reported no contacts, I proceeded down the still-cold ladder and took my submerged station on the diving stand.

Since I knew my diving station was on the stern planes, I sat in that outboard seat. To my left the chief of the watch sat at the ballast control panel (BCP) that contained electronic switches that controlled the ballast tank valves that allow seawater to enter the ballast tanks located between the superstructure and the pressure hull.

The BCP, with electronics and hydraulics, controls valves and pumps that move water to and from internal tanks. This process creates submerged trim (balance and stability) of the boat.

The diving officer (DO) took his seat on an elevated chair behind me. In front of me was the panel that showed the angle of the stern planes that along with the bow planes control the angle of the ship when underwater. My back-lighted instrumentation included a depth gauge, a compass showing ship's course, and a ship's speed indicator in knots.

The boat was sealed and the board was green on the BCP, indicating that all openings to the sea were closed. Although I was not a qualified submariner, I had learned how to man the diving station in submarine school. But I needed to demonstrate skill and knowledge at sea in order to get a sign off in my qualification notebook.

For the second time in my career, I heard the diving klaxon. Hair stood up on the back of my neck. The DO ordered a 10 degree down bubble, and I pushed the wheel forward slightly, but not too far, to accurately put the needle on 10 degrees down. Meanwhile I could hear the hissing sound of air being released from the ballast tanks. I braced my feet as *Seawolf* took her down angle. We were ordered to level off at periscope depth, so I eased back to zero bubble as we approached 60 feet.

Submarines don't submerge because they get heavier; they dive because they displace less volume of water when the ballast tanks are full, and the ship achieves negative or neutral buoyancy.

With frequent ordered changes in course, speed, and depth, I was gaining experience manipulating the controls as we went deeper and faster to reach our operations area. During the four-hour watch, I was moved to the bow planes and the helm where I developed more skill and confidence. I became very good at operating the depth control planes and the helm. It was satisfying to respond quickly to DO commands and smartly ease the boat into a new position.

Running submerged was comfortable except for the several times we went up to periscope depth. We did some rocking and rolling.

In addition to seaman duties: lookout, helmsman, and diving planes watches, I spent time with the quartermaster of the watch (QMOW) to learn navigation knowledge and earn sign-offs on the QM practical factors. My days at sea were busy and long.

Four-hour watches on the diving station and then another four hours as apprentice to the quartermaster of the watch, where I learned how to determine and maintain ship's position by dead reckoning using the dead reckoning computer (DRC) and LORAN. We copied LORAN signals while at periscope depth. LORAN (long range navigation) is a system of navigation using radio signals that was developed during WWII. *Seawolf* was not equipped with a satellite navigation system.

After we determined our geographic position, with a navigational fix, from either LORAN or the DRC, we marked the spot with a pencil on the chart along with estimated time of arrival (ETAs) at future positions based on our course and speed. After the next navigational fix, we would see how well we were dead reckoning. Much of the ocean has been surveyed and the ocean depths are listed on our navigational charts. We used bottom contour (sea depths), as well as a check, but did not rely on it for a valid ship's position.

My chief, Quartermaster Chief Jackson, was an alcoholic, as many sailors were. But at sea, he was an excellent teacher. He taught me much about navigation: my new trade. My chief earned my gratitude and respect because he was willing to mentor a two-year reservist.

Off watch was devoted to submarine qualifications (earning my silver dolphins) and becoming a real submariner.

I adjusted quickly to the routine of living and working in a steel cylinder, with white lights during the day and subdued red lighting at night, and never seeing the sky, sun, or clouds. By and large I enjoyed the feeling of submarine life, the often-cramped quarters and the adventure of being (comfortably) submerged at sea.

My submarine qualification progress was monitored. If I fell behind, I was designated as delinquent. I didn't need to wear a scarlet D on my blue dungaree shirt. It was a small ship, and everyone knew when you were delinquent. And when I was, I was banned from watching movies in the crew's mess, taking away the main source of mental therapy.

With the help of many shipmates and my chief, I was making good progress. I passed my written examination for petty office third class (paygrade E4) and was given a promotion date. In early 1965, I would sew the crow and single chevron patch onto my uniform sleeves.

With hazardous duty (submarine) pay I would then make $350 per month, the most money in my life at that time. That was about $12 per day. Since I was working sixteen-hour days, I would earn about 75 cents per hour.

Most of *Seawolf's* time at sea in 1964/65 was dedicated to local East Coast submarine exercises. We got underway routinely as *Seawolf* acted as a target submarine for the newer class of fast attacks, designated as the Permit class since the original lead ship in the family, *Thresher,* was lost.

Several times while on watch, I heard the high-pitched sound of an exercise torpedo outside our hull. One scraped down our starboard side from bow to stern. *Seawolf* made a good target ship and provided extensive torpedo-firing practice since we put out lots of noise into the ocean.

Barfly

Barclay was an electrician's mate. However, since nearly everyone on *Seawolf* had a nickname, he was called "Barfly." My nickname was "Franny." I thoroughly disliked it. Barfly was very helpful to me in my qualification. He took the time to help me learn *Seawolf's* complex electrical systems. It's been over fifty years, but I can still vividly remember Barfly. Whenever he was up and walking through the boat, he sang this song:

Jesus puts his money in the Chase Manhattan Bank,
Jesus Saves,
Jesus Saves,
Jesus Saves.

On one operation, I made my first voyage into the mysterious Bermuda Triangle and returned safely. This triangular section of the Atlantic Ocean is roughly bounded by Miami, Bermuda, and Puerto Rico, where dozens of ships and airplanes have disappeared. However, shipwrecks in this area are not any more common than in other parts of the world's oceans.

During that underway, we pulled into Cape Canaveral, Florida. The warm air felt great. After completing my topside watch I went into Cocoa Beach. In 1965 Cocoa Beach had not yet become a boom town. The mercury program was complete and the US was in early NASA Gemini two-man missions.

In my white uniform, I walked into a bar and sat down. But this time, no one offered to buy me a drink. After ordering a beer, I asked the bartender why there was a telephone and a number by each bar stool.

He answered, "If you see someone you like, give them a call. The astronauts do that."

The Mess Cook

My feelings of progress and fulfillment suddenly came to an end one day when the chief of the boat approached me, saying, "Francis, I need to send you mess cooking."

Mess cooking aboard a ship is probably the hardest and most-despised assignment, and I was overcome with dread. I asked the chief, "Why?"

He said he needed to do it before I became a petty officer, or he would be accused of favoritism.

"I'm okay with that, Chief."

"Sorry, Francis. You need to report to the cook tomorrow."

My quartermaster chief already knew. Of course, they had talked about it earlier and agreed to it in the "Goat Locker" (chief petty officer's quarters). It's a small ship. As ordered, I reported for duty to the senior cook the next day.

While I was a submarine mess cook, the mess deck became cold and sterile, not a place of enjoyment, more like a prison. I could not have been more unhappy. It's dirty work, but someone has to do it. *But why me, Lord?*

Each morning I arose at 5:00 a.m. and reported to the crew's mess to help prepare for breakfast; receive orders from the cook; get dishes, silverware, and serving platters from the lockers; set the tables; and bring the required food and beverages out of the storeroom, freezer, and refrigerator.

Most items (stores) were kept below the mess deck plates, which required climbing down and back up, carrying as many items as I could manage. Crawling into and searching in the freezer was the most physically difficult and uncomfortable task. It was cold.

Often I worked in the cramped scullery, cutting fruit and vegetables and preparing salads. When preparation was complete, under time stress and occasional verbal abuse, I became a waiter to serve and keep the serving dishes full. Having successfully fed the ongoing watch section, we turned around to feed the off-going watch: generally feeding a third of the crew, about forty sailors.

Ice cream night was the worst. Under time pressure and frequent shouting, I prepared dozens of soft-serve ice cream orders with numerous toppings like chocolate, cherries, sprinkles, and nuts. Many times I got the order wrong. In retrospect, I should have carried a note pad to record each order. But unfortunately, I didn't think of it.

Off to the scullery to wash the dishes, pots and pans, and silverware, and place the garbage into mesh bags. But first I needed to put a lead weight on the bottom of each garbage bag. Eventually, I sealed them and loaded them into the trash disposal unit (TDU).

The TDU is a small vertical torpedo like tube in the scullery, which penetrates through the submarine's pressure hull. Putting on my sound-powered headphones, I coordinated with the officer of the deck in the control room. "Request permission to dump garbage."

It was a smelly, dirty, controlled operation. Creating an opening to the sea on a submerged submarine is a careful and serious operation. Also, if the ship is involved in quiet operations or war games, it's not a good time to dump garbage.

When granted permission, I opened the inner door, loaded the containers, then shut and secured the inner door. When I opened the outer door, the weighted bags dropped to the bottom of the ocean where they would remain for all time (we believed.) And we would remain undetected because we left no trail of garbage.

At sea, this demanding mess cook activity went on, with a few breaks, for eighteen hours each day: breakfast, lunch, dinner, and two MIDRATS (midnight rations), providing soup and sandwiches for the off-going watch section. Five meals each day. When I finally hit my rack around 2300, I was dogged tired

Well, I'm ashamed to admit that I wrote some whiny letters back home to Linda. In one of her responses, she wrote, "A spoon full of sugar helps the medicine go down."

I didn't have the faintest idea where that came from, and I took no solace from it. Linda had just seen the newly released movie, *Mary Poppins*. The spoonful of sugar advice came from the lyrics of one of the movie's songs. Perhaps I complained too much. But I tell you, I was really unhappy.

I was a mess cook for three months. On top of the physical exertion and daily mental stress, I fell behind in my submarine qualification goals. However, I did learn how to make salads for forty people. I even learned how to make a great Waldorf salad. And that leads me to my rags-to-riches story. Yes. My wife, Linda, is still waiting for the riches.

The Squadron Commander

We were at sea, and I was in the scullery making a large, delicious Waldorf salad when the squadron commander, who was "riding" *Seawolf*, stepped into the door to speak with me. He was a full US Navy captain in charge of Submarine Squadron Ten, at State Pier in New London.

He said, "Francis, did you know you have the highest test scores of all of the sailors on this submarine?"

"No, Sir, I didn't." I responded.

He continued, "You do. And I want you to go see the executive officer and apply for the NESEP Program."

I promised that I would.

He pointed his finger at me and said, "Be sure that you do!" Then he departed.

Being aware of the Navy Enlisted Scientific Education Program (NESEP), I had planned to apply after I advanced to petty officer, thinking that would increase my chances of being accepted. Well, I guess I need not wait any longer. NESEP was designed to give "promising" enlisted sailors an opportunity to earn a college degree in either science or engineering. Apparently, the US Navy, at the time, had a shortage of technically trained officers.

During the 1960s, Submarine Squadron Ten, an elite squadron, was composed entirely of nuclear-powered submarines. It's flagship, the submarine tender *Fulton,* was moored at State Pier, New London. Submarine Squadron Ten's motto was "Find 'em, chase 'em, sink 'em."

How in the world, or why in the world, that squadron commander had knowledge of me and took an interest in my career, I do not know. I did not ask at the time, or even afterward. But I am forever grateful that he did.

As directed, I met with *Seawolf*'s executive officer (XO), Commander Phoenix, and applied for NESEP. The XO was encouraging and helpful. I was buoyed by a new vision of my future. One where I would eventually get off mess cooking duty, advance to petty officer third class, get assigned an actual bunk, earn my silver dolphins, and enter the queue for a college scholarship and commissioning opportunity.

One day there might be an Ensign Francis. That would make me the first in my family to earn a college degree and serve as a military officer.

Best of all, I learned that *Seawolf* was scheduled for a long overhaul at Portsmouth Naval Shipyard starting in May 1965. I would be ashore and make enough money to bring Linda out east.

After my tour of mess cooking was over, I gratefully returned to the Quartermaster Division as a newly promoted petty officer. On December 22, *Seawolf* was in port on my twenty-first birthday, and the QM gang agreed to celebrate with me. New London has many historic sites. But we were only interested in the bars and restaurants on State Street and Bank Street. We began at 8:00 a.m. with a shot and a beer. That gave me an early opportunity to show off my ID card.

Some breakfast, then on to our mission of bar hopping. Many of my drinks were purchased for me. We stopped for dinner, then continued upholding our reputations as drunken sailors.

After dark, I became ill and took a cab back to the pier. I managed to safely climb down *Seawolf's* forward hatch and ladder, then passed out in my bunk (I actually had a bunk then) still wearing my beer and vomit-soaked blue uniform. I survived and recovered from my epic, amateur performance.

Seawolf got underway in May 1965 bound for Portsmouth, New Hampshire. On the way into port, we stopped to hover, quietly and respectfully, over the site of the *Thresher* sinking. We switched all lights to red and maintained a reverent silence throughout the boat. The captain read from scripture: Genesis 1: "And God said, 'Let there be a vault between the waters to separate water from water.' So God made the vault and separated the water under the vault from the water above it. And it was so. God called the vault 'sky.' And there was evening, and there was morning—the second day. And God said, 'Let the water under the sky be gathered to one place, and let dry ground appear.' And it was so. God called the dry ground 'land,' and the gathered waters he called 'seas.' And God saw that it was good." (NIV)

It was a moving experience. 8,400 feet below us was the wreckage of one of our newest submarines with 129 souls aboard who died on a mission to advance submarine technology in defense of our country.

What did I learn?

- ✓ Persistence pays.
- ✓ Plan the work, and work the plan.
- ✓ Hope and hard work can guide us out of hopelessness and self-doubt.
- ✓ Overcoming adversity can be a decision. It can be a choice.
- ✓ Adapt to circumstances to make progress.
- ✓ Ship. Shipmates. Self. In that order.

CHAPTER 9
PORTSMOUTH ON THE PISCATAQUA

USS *Barracuda* in drydock at Portsmouth Naval Shipyard, drawing by Vernon Howe Bailey, 1941, courtesy NHHC, accession #88-165-CB.

Portsmouth, New Hampshire, sits across the Piscataqua River from Kittery, Maine, on the Atlantic Coast. New Hampshire's oldest settlement, and only seaport was a bustling Colonial port named for Portsmouth, England. The Portsmouth Naval Shipyard (in Kittery), dating from the 1790s, was an important factor in the city's economic growth. Portsmouth was a center for the submarine building and repair for most of the twentieth century.

Over the course of WWII, Portsmouth built more than seventy submarines, with employment swelling to over 25,000. Submarines have only been repaired there since 1971.

On May 5, 1965, *Seawolf* entered Portsmouth for refueling and an extensive overhaul, bringing her up to the SUBSAFE standard put in place after the loss of *Thresher*. After we tied up, the yard workers cautioned us to be ready to "pay out" lots of mooring line because of the high tides in the Gulf of Maine, some of the highest in the world. They were right. *Seawolf* rose high in the water at high tide then sunk well below the pier at low tide. Our constant line-handling duties were eliminated after we mercifully pulled into dry dock.

I was nearing the end of my submarine qualifications efforts. The final tests were a qualification board and a walk-through with a ship's officer, answering multiple questions about systems and operations. Fortunately, we completed the walk-through before the yard workers began to rip apart the ship. I was awarded my silver dolphins!

Submarine qualification, earning my silver dolphins, was by far one of the most difficult and rewarding challenges I ever faced. In keeping with an old and risky submarine tradition, we went into town and found a bar where I would drink my dolphins. This is probably similar to college fraternity hazing ceremonies. My shipmates filled a beer pitcher with several varieties of alcohol then dropped my shiny new dolphins to the bottom. My task was to drink the toxic mix of booze and end up with my dolphins in my teeth.

Being fearful of getting puking drunk, or even dying, I cheated by allowing most of the liquid to spill down my chin and over my uniform. In spite of several loud objections, catching the pin between my teeth put an end to the "ancient" ritual. I felt an extraordinary sense of relief and accomplishment.

Months of hard physical and mental labor, persistence, study, anxiety, stress, self-doubt, and help and encouragement from many shipmates, resulted in one happy sailor, soaked in booze, smiling, and biting a piece of silver metal between my teeth.

After my recovery, I turned my attention with excitement to making plans to bring Linda to New Hampshire. Several *Seawolf* sailors had found a neighborhood of bungalows in town. They were remodeled former US Air Force housing units near Pease Air Force Base. I went out to look and was able to reserve a cozy furnished one-bedroom, with bath, kitchen, and living room.

At home in Michigan, Linda had made amazing preparations for her move out East. She had purchased a newer car from her friend at the library. Yes. The 1958 Chevy Impala had been gently driven by a librarian.

Linda had a AAA road route plan ready. On her basement floor were several boxes packed with our home goods all stacked and arranged per the dimensions of a U Haul trailer, which she rented. I was impressed and elated. Our furniture would remain in Michigan until we could send for it later.

Our road trip to New Hampshire was like a honeymoon. We even stopped at Niagara Falls. Driving east, we saw many parts of the US we had never seen before. In Portsmouth, she loved our small but cozy bungalow and set about decorating and meeting our neighbors including several of my shipmates. Linda had saved some money which, in addition to my slightly increased military pay, meant that we could finally afford to be together. And we'd be livin' on love. Linda and I were overjoyed to be together again. With the exception of a few weeks leave in Michigan, we had been separated for almost a year.

Being ashore with limited duties allowed us to play Midwestern tourists, do some sightseeing, eat some sea food, and attend movies at the base theater. We drove to the White Mountains to see The Old Man of the Mountain and Mount Washington (6,288 feet). We didn't climb it, we just looked at it. We drove up to York Beach in Maine where Linda and I competed to see who could stay longest in the cold waters of the Gulf of Maine. We spent a weekend in Boston. Her first time, my second visit. We both fell in love with New England. Many friends came out to see us and explore the area with Linda. But the idyllic time came to an end after only three months.

As you recall, Navy quartermasters (ship navigation assistants) are pretty much a seagoing specialty. *Seawolf* had eight QMs, far too many for a ship sitting in the shipyard, and I was the most junior. In addition, it was highly unlikely that I, as a two-year reservist, would return to sea with the ship after her overhaul. Consequently, one day, I was notified by *Seawolf's* executive officer that I would be transferred back to New London to an operating submarine.

What did I learn?

- ✓ Goal setting focuses your attention and action.
- ✓ After some time away, homecomings can be sweet and uplifting. Almost, but not enough to make up for the separation.
- ✓ Focusing attention and effort on another person's happiness can be rewarding and promotes a healthy relationship.

CHAPTER 10

BANG

USS *Bang* (SS-385) alongside USS *Chivo* (SS-341) at San Juan Naval Station, Puerto Rico, during Operation Springboard; Canadian Ojibwa class submarine in the foreground. Photograph by PHC CJ Wiitala, USN, courtesy NHHC, catalog #NH 98698

Back at our bungalow, I told Linda I was being transferred back to New London. She was sad to leave Portsmouth. So was I. But at least we could remain in New England. Ever resourceful, she suggested driving down to New London to look for an apartment. That weekend, we did.

New London, Again

In New London, we bought copies of local newspapers and searched the apartment rental sections. An apartment on Montauk Avenue looked promising. The area had large, stately homes, many Colonial revival style. Some were reconfigured into apartment buildings.

We found street parking and knocked on a Montauk Avenue door listed as a rental. An elderly man opened the door. Linda said, “We are interested in the apartment.”

He said, “Okay. I’ll get the boss.”

An elderly woman, with bright-red hair, came out to greet us. She invited us to sit on the front porch. We introduced ourselves and asked if we could see the apartment.

She began interviewing us, “How long have you been married?”

Linda responded, “Three years.”

“Do you have any children?” she asked.

“No,” Linda said.

The woman leaned toward Linda and asked, “What’s the matter, you on strike?”

Next she asked me what I did for a living.

“I’m a sailor stationed on a submarine in New London.”

The conversation ended abruptly.

“The apartment is not available,” she said.

We were in shock. Was this blatant discrimination against a young military family?

We moved on. The next promising rental was in Norwich, "The Rose of New England." We spent the night there at the historic Wauregan Hotel, built in 1855. The room across the hall from us had a brass plaque on the door: "Abraham Lincoln slept here in 1860."

The upstairs apartment we found the next day on Wawecus Hill Road, nestled in woods, was gorgeous and affordable. The owners accepted our first-month's rent deposit.

We drove back to Portsmouth victoriously. We said goodbye to our friends, neighbors, and shipmates, and packed a U-Haul trailer and moved for the second time that year. Linda arranged for a moving van company to pick up our furniture stored in Michigan and deliver it to our new apartment in Norwich. Linda and I were happy to live in this lovely part of the country. Our new apartment was spacious, with two bedrooms, hardwood floors, and a wood-burning fireplace.

We scored on this one.

Diesel Boat

My new orders assigned me to the USS *Bang*, moored at State Pier in New London. *Bang* was a Balao Class diesel-electric submarine similar to the USS *Clamagore* that I sailed on for a day during submarine school. Built in 1943, *Bang* was small compared to nuclear boats like *Seawolf*. She was 331 feet long, displaced 1,550 tons, carried a crew of about seventy men, and was powered by diesel engines and large batteries when submerged. *Bang* had ten torpedo tubes: six tubes in the forward torpedo room and four tubes aft.

Prior to reporting onboard, I ironed and starched my summer white uniform, put it on, and kissed Linda goodbye. After I parked at State Pier, I went aboard the black vessel with the USS *Bang* sign hanging on the brow. The smell of stale air and diesel fumes immediately filled my nostrils. I reported to the ship's executive officer who was sitting in the small officer's wardroom, drinking coffee.

Inadvertently, I had overlooked a small detail. Believing that I presented the most impeccable appearance in the fleet, I felt deflated when *Bang's* executive officer ordered me to button my shirt pocket, which contained my ballpoint pen.

Petty Officer, QM-3 (SS) Francis in his "impeccable" uniform. (Photograph taken by Linda)

Reviewing my service jacket, he went on to inform me that I needed his approval before I would be accepted into NESEP. It was an uncomfortable and unwelcome "Welcome aboard." *Bang's* executive officer was also her navigator, so he would be working closely with me. Since I was a third-class quartermaster and submarine qualified, I would work in that division.

Next I met with the chief quartermaster who was aware that I was probably not going to be onboard for long.

Like *Seawolf*, *Bang* had just returned from a long deployment in the Mediterranean. I did not feel disappointed that I had missed several weeks of crossing the Atlantic Ocean on the surface. The chief of the boat assigned me a pull-down bunk in the after battery, near the crew's mess. The after battery is the crew's berthing and dining space and sits above half of the ships' electric batteries. The other half of the batteries are forward in an area appropriately named the forward battery. That's also where officer's quarters are located.

We would get underway in a few days for local daily OPS (operations), acting as a Soviet submarine in service to surface ships and other submarines conducting anti-submarine warfare (ASW) exercises. That made sense since the Soviet Union's large submarine force in 1965 was comprised predominantly of diesel-electric boats. (More opportunities to be hit by exercise torpedoes and dummy depth charges.)

Before we got underway, I took some time to read about *Bang's* history and her exploits on six WWII war patrols in the Pacific. *Bang* was a distinguished veteran of actual combat service. First, she was named after a fish, the bang, a dark-blue or black fish of the Atlantic herring family found in US coastal waters north of North Carolina.

Bang was built as a WWII fleet submarine at Kittery, Maine, by the Portsmouth Naval Shipyard. She was launched August 30, 1943, and commissioned on December 4, 1943 (The year I was born).

Bang was sent to Pearl Harbor. She made six war patrols in the Pacific, sank more than ten Japanese merchant ships, and survived more than 200 Japanese Navy depth charge hits to her hull. Assigned lifeguard station northeast of Formosa in support of the Allied struggle for Okinawa, *Bang* rescued a Navy pilot who ditched his plane after it was damaged by flak.

On October 4, 1952, following her post-war overhaul at the Portsmouth Naval Shipyard, *Bang* was recommissioned as a GUPPY IIA submarine. Her hull was streamlined, and additional power was added to the engineering plant to provide increased submerged speed.

Bang was ordered to the Caribbean Sea in fall 1965 and again in spring 1966 to participate in NATO anti-submarine warfare exercises, named *Springboard* and *Fallboard.* We would provide services for the US, British, and Canadian navies. *Bang* acted as an attacking Soviet diesel-electric submarine attempting to penetrate the defensive screen to make a torpedo run on the capital ship, often a US aircraft carrier.

Between us and the target was the "screen" comprised of numerous escorts, helicopters, and fixed-wing ASW aircraft (P-3 Orions), all looking for us with sonar, dipping sonar, sonobuoys, and magnetic anomaly detectors. If we snorkeled, running our diesel engines, they could detect us with a diesel exhaust sniffer.

The Canadians were especially good at coordinated ASW. While submerged, I was on watch in the conning tower when I heard a loud thud on the bridge.

Word came over the underwater telephone that we had been sunk and should surface. First to the bridge, I found an exercise depth charge (a cement warhead loaded hedgehog) rolling back and forth on the bridge deck. The Canadian aircraft had put one "right down our pipe."

When the captain arrived on the bridge, I asked if I could keep it as a souvenir. He said, "No. I'm sorry, Francis. We have to give it back to the Canadians."

The NATO wargames were exhilarating, and exhausting.

Rigged for silent running on one approach, the air conditioning and fans were shut off. In the warm Caribbean waters, the hull temperature reached 110 degrees Fahrenheit. It was hot, smelly, and stuffy. Trapped in a closed atmosphere inside a diesel electric submarine, diesel boat sailors constantly inhale diesel fuel vapors. My body, mind, and lungs were in distress.

Surfacing and opening the hatch brought full body relief as I felt and inhaled the cool, clean, tropical air. The blissful feeling was callously interrupted by a US destroyer sending us a flashing light message. And he was sending it rapidly.

My morse code was never very good. I was able to catch a few words when the officer of the deck arrived and asked me what the message was.

Knowing we had just provided ASW services to the US Navy destroyer, I said, “He’s sending, “Thank you for your services. You are released.”

I returned a slow flashing light: “Roger Out.” .-. --- --. . .-. --- ..- - or something close. Mercifully, the destroyer’s flashing light operator stopped sending and the vessel turned away.

Many evenings *Bang* sat on the surface running her diesel engines to recharge her spent batteries. It allowed us to recharge “our” batteries as well. It was easy to feel the romance of the sea while resting in the calm Caribbean, kissed by warm tropical breezes, and embraced by a blanket of stars.

Several times we pulled into one of the US Virgin Islands for liberty ashore: Charlotte Amalie, St. Thomas, Christiansted, St. Croix, and St. John. The islands had the most beautiful white beach sand I’d ever seen. We spent many nights listening to steel band music and sampling tropical drinks. Yet again, as a young sailor of limited means, my time ashore was limited.

We were warned not to swim off the submarine in port due to the abundance of barracuda fish. So I never did. But my shipmates and I did swim at night (with the sharks, I suppose) off some beaches in Puerto Rico. When I was young and stupid, I was young and stupid.

During the six-week deployment, some of our time ashore was spent in San Juan, Puerto Rico. One evening I was assigned shore patrol duty in town. I was pleased that my partner on patrol was a hefty sailor nicknamed Tiny, who weighed over 200 pounds. Looking back at our brief training, I’m surprised that we were not trained in de-escalation techniques. Instead, we were only taught moves to disable a drunken sailor with our night sticks.

Fortunately, we encountered no trouble or disruptions that night.

In our small submarine crew, every sailor performed manual labor. I was frequently assigned cleaning duties. "A clean ship is a happy ship." A clean ship is also a healthier ship, especially one that is frequently sealed off from fresh air.

As a junior petty officer, I was tasked with "good order and cleanliness" of my workspace: the conning tower. I presented the space with pride and never failed an inspection. During stores load, my commonly assigned position was naturally in the tight spaces outboard the diesel engines to receive and stack the "groceries."

One day, the chief of the boat directed me to scrub the deck in the after battery, crew's mess. I worked on my hands and knees, using scouring powder and a "greenie" as we called the 3M Scotch-Brite scouring pads. The chief came in to inspect my results.

He said, "Francis. This is the cleanest I've ever seen this deck." He understood the power of positive feedback. To this day his compliment ranks near the top.

Shipmates

Surprisingly, my best friends onboard *Bang* were not fellow quartermasters. They were electrician's mates named Jim and Mike. Together, we spent many hours ashore in the Caribbean. We were all young and newly married. Back in New London, we gathered as couples: Mike and Linda, Jim and Carol, and David and Linda. It felt like family.

The three young woman created a comforting support system when we sailors were at sea, which was most of the time. It was the first of several adopted families that Linda would embrace.

In the summer of '65, shortly after I reported aboard *Bang,* we had learned that Linda was pregnant. It was planned and joyful news. But only twenty-two years old, I felt a little anxious.

How am I going to support and nurture a baby?

Coincidently, Mike's wife, Linda, was also pregnant with their first child.

My life was filled with exciting adventures and long workdays. While Linda spent much of her time alone in our apartment, I went on Caribbean "cruises." For us to get out of town together, she arranged a bus trip to New York City to attend the 1964/1965 World's Fair at Flushing Meadows in Queens. The lines were long, but the exhibits were extraordinary.

While I performed well and was competent in my duties, I was regarded by some as an officer wannabee, a reservist, and a college kid—not one of the regulars. In addition, the executive officer/navigator (XO) continued to hold his NESEP recommendation over my head and his approval that I needed to transfer off the ship and into the program.

On watch one day, I mistakenly wrote, "senoir officer present afloat" (SOPA) rather than senior officer. I was subjected to intense, gleeful ridicule by my chief. "College boy doesn't know how to spell."

Creating nicknames for others, some of them cruel, is as common as saluting in the Navy. Many in the crew referred to the XO as "Hairy Finger." He received that nickname because he was missing a fingernail, and dark hair grew on that bare finger.

After returning from Springboard, *Bang* resumed a pattern of daily operations off the East Coast. At sea, anchored off Block Island for the evening, I had the anchor watch and was stationed in the conning tower. Using the periscope, I was plotting our position regularly, using three structures on the island, when I noticed we were moving. We were dragging our sea anchor on the ocean bottom and our ship was being pushed east by a strong wind.

I called the navigator (Hairy Finger) to the conn. He was outright skeptical. "Are you sure? Are you sure?"

Pointing to the chart, I directed his attention to my plots of ship's position. The numerous moving and enlarging triangles on the chart indicated that we were drifting east.

"Well done, Francis," he said.

"I'll set the maneuvering watch"

As he climbed down the ladder, he yelled, "You've got my recommendation."

At last! I thought.

As the wife of a seagoing sailor, Linda proved to be what was commonly referred to and prized as "a good Navy wife." This is not a title of condescension, but of high praise. She kept the home fires burning. She was amazing at managing the house, and money, and tolerating a returning, grumpy, smelly, submarine sailor.

Before I departed *Bang,* I had a near accident. We were on the surface, in the North Atlantic, in rough seas. As I was climbing down the steel ladder from the bridge, *Bang* rolled heavily to port. I lost my footing. Grasping the ladder with only my hands, my body swung out 90 degrees to the ladder. As *Bang* rolled back to starboard, my whole body slammed into the ladder. I lost my grip and fell down through the submarine's tall sail structure and slipped through the Conning Tower. As I was falling through the Control Room hatch I was able to grab hold of the hatch coaming with my fingers, stopping a 25-foot, fast-moving fall onto a steel deck below.

I looked like a circus act. I was shaken and further insulted by the laughter from everyone in the watch section. But I escaped serious injury.

It really happens! I experienced the "my life flashing before my eyes" phenomenon as I was falling from the bridge. Later, I did some research and found evidence of this in medical journals. The medical term is *compressed life review* (CLR), a mnemonic illusion of having "your entire life flash before your eyes." This was my first time. But it would not be my last.

What did I learn?

- ✓ Diligence, hard work, and good performance will often be recognized and rewarded.
- ✓ Good friends are important for our happiness, well-being, and mental health.
- ✓ Serving aboard a submarine that saw combat heightened my sense of appreciation for our WWII veterans. Two combat veterans of WWII, who had made war patrols and had experienced enemy depth charging, were still serving aboard *Bang*. They were both alcoholics. Alcohol was their drug of choice for coping with their "shell shock." Now commonly understood as post-traumatic stress disorder (PTSD) and Traumatic Brain Injury (TBI).
- ✓ PTSD/TBI affects and burdens thousands of US combat veterans from our many wars and conflicts.
- ✓ When I read about the war patrols of our US submariners during WWII, I think about Winston Churchill's Finest Hour Speech in 1940: "Never in the field of human conflict was so much owed by so many to so few."

CHAPTER 11
AN OFFICER CANDIDATE AND A FATHER

The path was now open. My opportunity to earn a college scholarship and a Navy commission was a reality. No more hurdles or roadblocks. Well maybe just one. Linda wanted me to get out of the Navy and go back home to Bay City. She wished to be close to her family, especially with our first child.

However, leaving the Navy would mean not fulfilling my long-held, boyhood dream of becoming a nuclear submarine officer. So I moved forward, signed a reenlistment contract, and accepted my earned promotion to petty officer second class (Paygrade E5), QM2 (SS) Francis, United States Navy, and I received a small reenlistment bonus. After five years as a reservist, I was now a regular Navy man.

Things were tense for a while. In addition to our close Navy friends, we were blessed with amazing neighbors, Raymond and Edna, the young parents of three children, Cindy, Robert, and Gregory. They were life coaches to us. Raymond was a Navy veteran. In fact, he had served on the USS *Constitution*. Edna helped Linda work through her disappointment. You could call it tough love. While Linda was lamenting my decision to remain in the Navy, Edna said to her, "You did marry a sailor, didn't you?"

Some of her disappointment gave way to anticipation of the birth of our first child. Linda and I went shopping for baby furniture for the nursery. Our second bedroom in our apartment would become our nursery.

While selecting a crib, I noticed a contest display: “Guess the date of the birth and win a $100 coupon.” I filled it out and guessed May 29, 1966. May 29 was John Kennedy’s birthday and the date I enlisted in the Navy.

Robin Linn Francis

Linda and I were at home together, the evening of Saturday, May 28, when she went into labor. We drove to the Submarine Base Hospital in Groton. She was admitted, and I was told by a nurse to go home. They would call me in the morning. Apparently, they were anticipating several hours of labor. I returned to Norwich tired and went to bed. Around 4:00 a.m., on May 29, I was awakened by the phone. It was Linda. She said, “Congratulations. You are the father of a baby girl.”

In the early morning darkness, I returned to the Submarine Base Hospital in Groton to meet our firstborn. Visiting hours had not yet started. The nurse brought our daughter to the window of the double doors and showed our baby to me. Not until the next day, would I get to hold her.

We always knew our first daughter would be named Robin. Linda had picked out her name years before. As a young girl, Linda read Dale Evans’ book, *Angel Unaware*. She was moved and inspired by the story of Dale Evans and Roy Rogers’ daughter, Robin, who was born with Down syndrome and a heart defect. In the early ’50s, kids born with Down syndrome were usually placed in institutions, but Dale and Roy kept her at home, loving and caring for her.

The next morning, I was able to visit Linda and hold our first child for the first time. Linda was weary and had a saddle block (low spinal anesthesia) headache. I learned later that the medical staff ignored and never treated her headache, and it lasted for several days. When she told her doctor she had a headache, he said she didn't look like she had a headache, and he left.

In retrospect, I feel she did not get very good medical attention, and I was annoyed that I was unable to be with her during the birth. Remember, this is the 1960s; I suspected that the wife of a junior enlisted man was low on the staff's hierarchy of care. However, at the time, we were delighted with our healthy, beautiful, baby girl and a hospital bill of only twenty-five dollars.

Preoccupied with my success in the Navy, I was ill-prepared to be a father. I hadn't done my homework or assembled a survival guide. It would be trial and error, and I committed lots of errors. Along with feeling frightened and inadequate, I felt intense joy holding our baby daughter. Linda was right. She said, "David. Robin is the second woman you ever fell in love with."

Linda and Mike had a baby boy they named Michaelson. The two Lindas bonded closely because of their shared condition. In the summer of '66, my Linda, who grew up babysitting for her older sisters' children, was able to help Mike's Linda with baby care in the early months before we left Connecticut.

Just before Robin's birth, I received orders to report to the Naval Preparatory School in Bainbridge, Maryland, on June 1, 1966 for NESEP preparation. Linda was still in the hospital when I left her and Robin in the tender care of our neighbor Edna. Edna drove Linda and our baby back to Norwich. I hoped to return to Connecticut on a weekend as soon as possible.

Three days after Robin's birth, I made a lonely drive south to Maryland, all the time desperately wanting to be home with Linda and our newborn. But I was familiar with official orders taking precedent over personal plans.

Well, this is what I wanted. This is what I'd been striving for.

Port Deposit

Naval Training Center, Bainbridge, is located on the shore of the Susquehanna River near Port Deposit, Maryland. I checked into the barracks in Bainbridge and was assigned a room and a roommate—a US Marine.

The academics were familiar: science, math, and English. With the help of my prior education from Junior College, I did well. My roommate did not. He washed out, leaving me with a private room.

It was like summer school. Well it was summer. The curriculum was shorter but similar to the Naval Academy Preparatory School (NAPS) program designed to prepare midshipman for the US Naval Academy. Most of the instructors were university professors working a summer job. Most of my fellow students were white, and male. We were called NESEPers. The military loves acronyms. Many of my classmates were Nukes (nuclear power trained engineering rates). But I did meet one fellow quartermaster, QM1(SS) Bob, also a submariner.

I managed to return to Norwich a few weekends. But I was of limited help to Linda as she struggled to raise a baby, mostly by herself. Preoccupied with homework, road weary, and with a serious case of fear of failure, I was not very helpful.

A table in our large classroom at the training center was covered with college catalogues. After reading most of them, I chose to carry the Princeton University catalogue with me for my interview with the program director, Dr. Johnson.

He asked me what major I had selected.

I said, "Chemistry," knowing that the S in NESEP stands for scientific.

The Navy needs more officers with technical training. "We don't need any more officers with history majors," we were frequently advised.

"Could I go to Princeton?" I asked.

Dr. Johnson said, "No. We are sending chemistry majors to the University of Louisville."

"That's in Kentucky, right?"

"Yes," he responded.

He added, "It's one of the top ten chemistry schools in the country."

My fate was sealed in a brief, five-minute negotiation. I phoned Linda to tell her that we were on our way to Louisville. That's in Kentucky.

Later I received orders to report to the Naval Reserve Officer Training Center (NROTC) on the University of Louisville campus.

My new friend, QM1 (SS) Bob, was heading to Louisville as a chemistry major as well.

After graduation from Navy Preparatory School, I returned to Norwich to help Linda pack up again and watch our growing number of household goods loaded onto a moving van bound for Kentucky.

What did I learn?

- ✓ Children will change your life.
- ✓ Communications is the highest form of respect. It's respectful and good manners to discuss important life-changing decisions with your family or significant other. "Do as I say. Not as I did."
- ✓ "If a window of opportunity appears, don't pull down the shade." —Tom Peters, *In Search of Excellence*
- ✓ "In the middle of difficulty lies opportunity." —Albert Einstein.

CHAPTER 12

LOUISVILLE ON THE OHIO

In August, we drove west to Kentucky with our three-month-old daughter. In contrast to our honeymoon-like trip to New Hampshire, just over a year earlier, this trip was not as enjoyable. We were not tourists. We did no sightseeing. We stopped only to eat or sleep overnight. We were parents intent on a mission to establish a nest, a new home, to raise our child and build a better future.

Like veteran movers, we searched for a house to rent. Eventually, we found a great three-bedroom brick rambler in a suburb south of Louisville. I relied heavily on Linda's skills to organize the move and reestablish our family of three in a new home. She was really good at it.

Our first house! Although we were still renters, it was a wonderful feeling. We had lots of room, a basement, our own laundry facility, and a large backyard for Robin to play in. However, that meant that we needed to buy a lawnmower and cut the lawn for the first time. Linda also purchased a dishwasher on wheels which I regarded as one of the greatest inventions of mankind. And she went downtown Louisville to buy a beautiful walnut desk to replace my homework table.

After we were in our new house for a few days our neighbor next door came over to meet us. She introduced herself, "I'm Gigi."

Gigi looked at our baby in the basinet, looked at me, then looked again at Robin. Turning back to me she asked, "Are you the brother or the father?"

I said, "I'm the frightened, twenty-two-year-old father."

We were blessed once again with great neighbors: Gigi, George, her husband, and their two children, George, Jr. (Georgie) and Karen. George was a World War II Army Air Corp veteran who had flown bombing missions over Germany on B17s, as a tail gunner, surviving multiple missions.

The Mother of all Scholarships

After hearing it a few dozen times, we realized that Kentuckians pronounced Louisville, *Luvul.* It required rounding your mouth and using some rarely used facial muscles. After we settled in, I reported to the Naval Reserve Officers Training Corp (NROTC) Center and registered for the fall semester.

In addition to receiving full Navy pay and benefits, all college expenses were paid by the Federal Government, and we would be issued free books. The NROTC Center had a library full of college textbooks. Check them out and return them after completing the course. This was truly the *mother of all scholarships.*

However, there was one expense that was not covered under the *mother of all scholarships*: laboratory glassware breakage. As chemistry majors, we would, without a doubt, break plenty of glassware.

Being resourceful sailors, one of our NESEP upperclassmen removed a large tin of coffee from the NROTC and presented it to the Laboratory Stock Room supervisor. That expense was henceforth covered for my remaining three years at Louisville.

The prospect of being in one place for three years was strange and wonderful. We spent some of my reenlistment bonus money on a new car. We purchased a brand-new Volkswagen Beetle for $1600. Being a child of the '60s, I had always wanted to own a VW Beetle. We frequently sped back and forth to Michigan in our Bug for family visits during semester breaks.

By Navy standards, it was light duty. No long deployments. No standing underway watches on steel decks in tossing seas. No family separations. Other than wearing our uniforms on Wednesdays, alongside the NROTC students, we had no military obligations. My military responsibility was to earn a college degree.

At the same time, it was hard duty. Soon it became obvious that I had no business being a science major. Sadly, I lacked both the interest and the aptitude. It was hard stuff. If I had a choice, I would have been a history major. But as you recall, the Navy didn't want any more officers with history degrees.

I suffered through three semester of calculus, obscure subjects like physical chemistry and atomic physics. Much of the nuclear subatomic particle theories (like neutrinos) sounded like mysticism to me. Struggling to maintain a B-average, I took heart in the knowledge that chemistry majors always have a lower grade point average (GPA).

Despite my lack of aptitude, after a successful first year of study, I was confident that I could survive to graduation and maintain a "gentleman's B." I did.

Surprisingly, the Naval Ordnance Station in Louisville hired me to teach Chemistry 101. I enjoyed teaching and did pretty well while earning a phenomenal five dollars per hour.

It became apparent to me why Louisville was highly regarded. Founded in 1798, as the nation's first city-owned university, Louisville, or U of L, was a leading doctoral and research institution. At that time there were about 2,000 students on campus. Class sizes were small and nearly all of my courses were taught by a full professor (PhD).

Fortunately, I was able to take several, enjoyable liberal arts requirements and electives. And they raised my grade point average.

My freshman college English class was stimulating. One of our early assignments was to write a short story about a memorable personal experience. I wrote about *Seawolf,* my first submarine, and our time hovering over the site of the sunken *Thresher:*

> We got underway in May 1965 bound for Portsmouth, New Hampshire. On the way into port, *Seawolf* stopped and hovered quietly and respectfully over the site of the *Thresher* sinking. We switched all lights to red and maintained a reverent silence throughout the boat . . . It was a moving experience; 8,400 feet below us were pieces of one of our newest submarines with 129 souls aboard who died on a mission to advance submarine technology in defense of our country.

One of my younger classmates asked me what I wrote about. I said, "I wrote a sea story."

He responded, "You mean you only wanted to get a C?"

"No, I got an A by writing a S-E-A story."

During one of my elective political science classes, we were assigned a research paper on an area of professional interest. At that time, the Soviet navy was arming its warships with surface-to-surface missiles that threatened our surface Navy ships and task forces—a known vulnerability of the US Navy. This military threat was frequently in the press.

The Navy's public response was "our carrier aircraft were the first line of defense."

I didn't buy that. There was not always an aircraft carrier in the vicinity of our ships. Many Navy ships sometimes traveled alone.

Before I graduated from Louisville, I was relieved to learn that our Navy had developed a simple, but elegant, solution: A gatling gun, as in a rapid-fire, deck-mounted, machine gun (the weapons system known as C-RAM).

Outside of class, and at home, life was much easier, thanks to Linda. I relished my time with our young daughter, giving her baths, reading to her, dancing and playing with her. It was our father/daughter bonding time, joyful and rewarding. There was hope for me.

Linda gave me ample time to do my homework. She managed the house and kept us afloat on my still-meager military pay. Losing hazardous duty (submarine) pay reduced our income. As a second-class petty officer (Paygrade E-5) with over four years of service, I made about $400/month. Linda made it stretch.

Once a month, Linda drove our VW to the commissary at Fort Knox, Kentucky Army Base. She filled the car with groceries that actually fed us for a month. Linda mentioned that many of the army personnel looked really poor. She felt we were actually well off compared to some Army enlisted.

Being ashore gave us more opportunity to have a social life. We joined an Episcopal Church nearby and made new friends. Similar to *Bang,* we enjoyed the camaraderie of several Navy couples (especially Bob and Carol, and Jim and Judy.) Along with the friendship and support of our neighbors and several other "civilian" friends., it felt like we were part of a community.

In keeping with her tradition of finding adventure and culture for us, Linda purchased a membership in the Actors Theatre of Louisville. We enjoyed professional theater at a modest price.

Senior classmen got us involved in extracurricular sports. Proudly wearing navy-blue and gold jerseys, we NESEP students, we Navy enlisted men, fielded a softball team in the Louisville City League. And we were out of our league. I was selected as the starting pitcher and ended my career with a perfect record: never winning a game.

Linda joined a campus group called the DAMES Club, a social support group for spouses of college students. She would eventually earn her PHT, Putting Hubby Through.

Celia

It was unexpected and unwelcome. Linda's mother, Celia, died of heart failure at just fifty-six years old. On leave from the university,

Linda and I and our young daughter drove up to Michigan. It was February, cold and overcast. She was the first of our parents to die. We were not ready for it.

Celia was kind to me from the time I met her, when I was a teenage boy dating her daughter. She was warm, funny, and loving. Her death was the first real experience we had with grief and loss. There would be others.

As a result of my mandatory annual physical examination in 1968, I was ordered to undergo an inguinal hernia repair. I checked into Ireland Army Hospital at Fort Knox for the procedure. After surgery, I shared a room with an army warrant officer, helicopter pilot, who had been shot down in Vietnam and severely injured. I thanked the warrant officer for his service and bravery and promised not to complain around him about my elective surgery pain.

In retrospect, the 1960s could well be regarded as the most consequential and turbulent decade in American history so far. While nestled in a small, quiet college and community, Linda and I religiously watched the evening news together. We witnessed massive social and political upheaval, rising demands for civil rights, and growing resistance and protests to the war in Vietnam. Two American leaders, while waging peace, were assassinated: Dr. Martin Luther King, Jr. and Senator Robert F. Kennedy.

And the US space program suffered a major tragedy. On January 27, 1967, NASA experienced its first space disaster: the deaths of three astronauts during a training exercise for the Apollo 1 mission. The three Apollo 1 crewmembers intended for the first manned Apollo space flight were Edward H. White II, Virgil I. "Gus" Grissom, and Roger B. Chaffee.

A fire inside the Apollo Command Module during a test took the lives of all three astronauts. NASA had not before experienced a disaster of this magnitude.

Disturbing 1968 News Headlines

North Vietnamese communists launched the Tet Offensive, an assault that contradicted the Johnson administration's claims that the communist forces are weak, and that US-backed South Vietnam was winning the war. The highest casualties of any year during the entire Vietnam War were in 1968, taking 16,899 American lives.[8]

After a battle for the Vietnamese village of Ben Tre on February 5, where American bombs, rockets, and napalm obliterated the town, an American officer told Associated Press reporter Peter Arnett, "It became necessary to destroy the town in order to save it." That quote was printed in newspapers nationwide and became a catchphrase for opponents of the Vietnam War.

On June 8 in Memphis, Reverend Dr. Martin Luther King Jr., in town to support the sanitation workers' strike, was fatally shot on the balcony of the Lorraine Motel. Gunman James Earl Ray, a White supremacist, fled the country. Riots ensued in more than 100 cities nationwide that week, leaving dozens dead, more than 2,600 injured and many thousands arrested.

---

[8] US Government National Archives. Vietnam War U.S. Military Casualty Statistics. https://www.archives.gov/research/military/vietnam-war/casualty-statistics

After midnight June 5, 1968, Senator Robert F. Kennedy was celebrating his win in the California presidential primary with a speech aimed at unity when he was assassinated at the Ambassador Hotel in Los Angeles. Gunman, Sirhan Sirhan, a Jordanian citizen of Palestinian descent, was captured at the scene.

That August at the Democratic National Convention in Chicago, police and Illinois National Guardsmen went on a rampage, clubbing and tear-gassing hundreds of antiwar demonstrators, news reporters and bystanders, with much of the violence broadcast on national TV.

During that turbulent summer, Linda became concerned when there were civil rights protests and some violence near the area in Louisville where we played softball. Protests against police mistreatment in Louisville turned violent, and a week of disturbances ended in the arrests of six African Americans–dubbed the "Black Six"–on charges that they conspired to blow up Ohio River oil refineries.

Many evenings, for many years, Linda and I watched news reports of Americans demonstrating and protesting for equal rights. And we heard them sing, over and over, the anthem of the civil rights movement: "We Shall Overcome."

Many times I longed to join the demonstrators. Similar to my feelings of regret for "sitting on the side lines" during the Cuban Missile Crisis, I felt frustration that I could not join the movement more openly for fear that being arrested and jailed for civil disobedience would damage my Navy career. And of course, the last thing I wanted was more days away from my family.

USS *Scorpion*

Scorpion during builders trials off New London, Conn. June 1960. Vice Adm. Hyman G. Rickover, head of Naval Nuclear Propulsion Program, stands on the sailplanes with another officer, courtesy NHHC, photo NH 97215.

On May 22, 1968, USS *Scorpion,* a US Navy Skipjack Class nuclear attack submarine was lost while returning to Norfolk after a three-month deployment to the Mediterranean Sea with ninety-nine crew members aboard—the second sinking of a US submarine in the 1960s. The submarine had originally been scheduled to sail straight home from the Mediterranean to Norfolk, but on May 17, was ordered more than 1,000 miles southwest, toward the Canary Islands off Africa's coast. The US Navy wanted to check out a group of Soviet navy warships, including at least one nuclear submarine, operating in the area. By May 28, the story of the missing submarine was on the front pages of US newspapers.

Scorpion's shattered hull was photographed in October by a camera mounted on an unmanned "sled" tethered to a three-mile-long cable towed by the research ship USNS *Mizar*, which had searched a 12-square-mile area for weeks southwest of the Azores where officials calculated the wreckage lay on the seabed two miles down. A Navy investigation and court of inquiry concluded that the certain cause of the loss of *Scorpion* could not be ascertained from the available evidence.

I was undeterred by the losses in my small, chosen seagoing community. Despite two US Navy submarine tragedies in five years, I was still eager to serve again aboard a nuclear boat. Linda knew this and accepted it.

Late in 1968, the news from outer space was more encouraging. The US was winning the space race as Apollo 8 became the first manned spacecraft to orbit the moon and return safely in December.

Meanwhile, at the University of Louisville, a number of Black students occupied the Administration Building and demanded courses in Black history. Their request was promptly granted, and they dispersed. I began to call the University of Louisville a "hotbed of rest." We Louisville Cardinals (Louisville's mascot) quietly went about our business of learning even as peaceful and productive civil rights marches and demonstrations occurred in segregated Kentucky and in Louisville.

Maintaining her tradition of seeking adventure and new places for us to see, Linda arranged a trip to Gatlinburg, Tennessee, in the Smoky Mountains. We visited Pigeon Forge long before Dollywood.

The great mountain beauty and lots of arts and crafts creators was just the kind of stuff Linda loved.

Jocelyn Ann Francis

On December 11, 1968, Linda gave birth to our second daughter. Jocelyn was an uncommon name back then. This time, Linda delivered in a civilian hospital, and I was present at the birth. For the first time, I witnessed the miracle of birth. Seeing a child of mine born was a beautiful and amazing experience. I now had the pleasure of bathing two daughters and carving out some time to be with them.

Our hands were full. I needed to provide more help in the limited time I had at home. But it was a challenge. It became even more apparent to me what an amazing woman I married, as Linda continued to keep it all together. On December 25, we marked our third Christmas in Louisville, first Christmas as a family of four, and the completion of my junior year. That year, in addition to all that happened in our country, our family had mourned a death and celebrated a new life.

What did I learn?

- ✓ Moral leadership, often in the face of danger, is critical to keep the promise of the American Experiment that "all men are created equal."
- ✓ "Everybody can be great . . . because anybody can serve. You don't have to have a college degree to serve. You don't have to make your subject and verb agree to serve. You only need a heart full of grace. A soul generated by love." —Dr. Martin Luther King, Jr.
- ✓ "There are those that look at things the way they are, and ask why? I dream of things that never were, and ask why not?" – Robert Kennedy quoting George Bernard Shaw.
- ✓ Never major in science if you can avoid it. (I'm kidding)

CHAPTER 13
SUMMER OF ’69

Leaving Linda and the girls was difficult. We had not endured a lengthy separation in over three years. Jocelyn was barely six months old when I flew off, in uniform, to Newport, Rhode Island, in the summer of ’69 to be part of a thirteen-week course.

Officer Candidate School (OCS) is one of four officer training schools at Naval Station Newport, designed to ensure a working knowledge of the Navy (afloat and ashore) and prepare candidates for a commission. To earn the honor of a commission in “the world’s greatest navy,” OCS tests one morally, mentally, and physically to determine if one will be able to lead sailors and conduct worldwide naval operations in defense of the US.

After arriving at the Newport airport, I was studying the bus schedule to Naval Station, Newport, when a young naval officer and his wife approached me and asked if I needed a ride. They were headed to the base and offered to take me along. On the drive, a local radio station was playing, “Bad Moon Rising” written by John Fogerty and performed by Creedence Clearwater Revival. I commented to the couple that for a few months I thought the lyrics were, “Bathroom on the right.” They both laughed and said they thought so too.

They dropped me off, and I thanked the young couple for their kindness and checked into the barracks. My new roommate was already occupying our small, austere room containing two single beds, two closets, and two desks. Jeff was a native of Chicopee, Massachusetts, and a NESEP student at North Carolina State. He was also a surface sailor.

Most of my seagoing career, I had served primarily with submarine sailors. This would be my first long term exposure to a "surface" type. After several days of kidding each other, I began to earnestly try to understand his preference for the surface Navy—those ships of the fleet we submariners called targets. Jeff would later become my most faithful sailing partner.

All of my fellow Naval Officer Candidate School classmates were NESEP students. It appeared that we were being separated and segregated from the other non-enlisted student classes so that we would not contaminate their attitudes about the "Mickey Mouse" military requirements.

But the Navy school did its best to enforce a limited degree of discipline on all of us "white hats." Although by then, we "white hats" (NESEP students) had been issued officer-type caps and double-breasted, blue officer-type uniforms with a single star on each sleeve. The star is the designator for a US Navy line officer. What was missing, and to be added later, was one gold ensign stripe.

Among my classmates were a few US Navy chief petty officers. They were naturally placed in leadership positions. Like boot camp, we marched together frequently, sometimes in the rain. On Saturday mornings we marched in a pass in review (just like boot camp) prior to our short weekend leave, which started at noon.

Friday evenings at OCS were devoted to athletics: physical training. I chose to play soccer because I had never played the game and wanted to learn. On the field (pitch), I demonstrated neither skill nor an understanding of the rules.

When we were released at noon on Saturday, I went down to the Navy yacht basin and learned how to sail a sailboat. I was better at sailing than soccer.

Linda and I had crewed on a sailboat with our friend, Si Fisher, in the Saginaw Bay (Lake Huron), so I was familiar with the sport. After many hours under instruction, I passed my written examination and then the "behind the wheel" test on the water. After earning my license and the freedom to "captain" my own Navy-owned boat, I frequently invited my roommate and other classmates to sail with me on weekends.

The summer brought mostly sunny and warm weekends, and the winds on Narragansett Bay were always favorable. Sailing was incredible mental therapy, stress release, and good seamanship training.

My classes and study were rigorous and sometimes challenging, especially the tactics course. Much of it was new knowledge for me.

OCS is designed to provide the baseline knowledge required of commissioned officers including instructions in division officer fundamentals, naval history, engineering and weapons, damage control, leadership and ethics, watch officer role, navigation, and naval tactics and seamanship. After years of struggling with science courses, I finally achieved some academic success by acing the navigation course (since that was my field).

We also went out to sea. Well, at least we went out into Narragansett Bay. Each of us got to "command," take the helm of a small motorboat, learning how to steam in formation and how to observe the "rules of the road." For me this was by far the most enjoyable training at OCS.

USS Buttercup

My past exposure in boot camp and submarine school prepared me for the feared and legendary Damage Control Wet Trainer. Working within a team, in a noisy, chaotic and tight space you use hand tools, pipe patches, timbers and even mattresses to stop the flooding and prevent the USS Buttercup from capsizing and sinking. We survived.

Fortunately, the Apollo 11 moon landing took place on a weekend. I was able to watch it on TV in a bar in downtown Newport. We celebrated winning the space race. But the Cold War and the war in Vietnam still raged on.

The primary objective of Apollo 11 was to complete the goal set by President Kennedy in 1961: to perform a crewed lunar landing and return to Earth. Apollo 11 launched from Cape Kennedy on July 16, 1969, with Commander Neil Armstrong, Command Module Pilot Michael Collins, and Lunar Module Pilot Edwin "Buzz" Aldrin. On Sunday, July 20, 1969, hundreds of millions watched Armstrong step onto the moon and heard him describe taking "one small step for a man, one giant leap for mankind."

The three Apollo 11 astronauts returned safely to Earth; the Command Module splashed down in the Pacific Ocean on July 24 and was retrieved by a crew of sailors and hoisted aboard the US aircraft carrier *Hornet*.

The Apollo 11 Command Module is hoisted aboard USS *Hornet* CVS-12, courtesy NHHC

After I participated in a formal graduation ceremony and a final pass in review, I hoped I would never have to march in formation again. Following graduation from OCS, I returned to Louisville. Happily, Linda met me at the airport with our two daughters.

When I attempted to hug nine-month-old, Jocelyn, she pulled away and started to cry. She had forgotten who I was. For the next several months I worked hard to re-bond with her and my family.

What did I learn?

- ✓ Military service, with frequent and often long family separations, strains the bonds of human relations. The difficult balancing act to combine service to your nation and service to your family requires commitment, skill, and dexterity.
- ✓ "Twenty years from now, you will be more disappointed by the things you didn't do than those you did. So throw off the bowlines. Sail away from safe harbor. Catch the wind in your sails. Explore. Dream. Discover." —Mark Twain.

CHAPTER 14
ENSIGN, UNITED STATES NAVY

Image courtesy of military-ranks.org.

Ensign (Paygrade 0–1) is the entry-level commissioned officer's rank in the US Navy, equivalent to the second lieutenant rank in the other armed services. Ensigns generally serve on naval vessels as division officers, leading a crew of seamen and petty officers in a specific division such as engineering or administration. As junior officers, ensigns engage in on-the-job training and receive mentoring and training from superior officers and experienced petty officers.

Fall classes for my final semester resumed at the University of Louisville. I would graduate midyear, in December 1969. My toughest courses were behind me. My academic confidence had grown so much that I even took two graduate school chemistry courses. I was filled with anticipation and a deep sense of achievement that I had survived and would soon earn a bachelor of science degree. The thought never left my mind that I was about to be the first in my family to graduate from college. (You may have heard this before.)

I was also scheduled in the fall to advance to petty officer first class (paygrade E6). I was still very successful in passing Navy tests, and my practical factors had been completed while onboard my two submarines.

Fortuitously, the quartermaster rate was “wide open” with a shortage of sailors in that specialty. So I was able to advance from seaman to petty officer first class in just four years. Soon I would gain the longest title I ever carried: Officer Candidate, Petty Officer First Class, Submarine Qualified Francis, abbreviated OCQM1 (SS) Francis.

In September, I was invited to meet with the NROTC professor of naval science. He greeted me and said, “Mr. Francis. The Navy needs carrier pilots. Would you be interested in going to naval flight school in Pensacola, Florida?”

There was a shortage of Navy carrier pilots in the 1960s because of the war in Vietnam. Many of our carrier pilots had been shot down over North Vietnam, killed, or were being held as prisoners of war. I had never wanted to be a fighter pilot, and my motion sickness would have been a real problem in the cockpit.

“No thank you, Sir.” I responded, “I would like to return to submarine duty.”

He said, “Okay. I’ll request orders for you to Officer Submarine School. I’ll also recommend you for nuclear power training.”

To advance, and to command a Navy nuclear-powered submarine, requires becoming a nuclear-trained officer. My application to Admiral Rickover's nuclear power program office was rejected. Apparently, he was not impressed with a former quartermaster with a chemistry degree and a B-average. However, I did receive orders to Officer Submarine School in New London/Groton, Connecticut. This would give me an opportunity to serve aboard a nuclear submarine as a division officer.

December 22, 1969

It was the best of all birthdays. On December 22, 1969, I turned twenty-six years old, earned my college degree, and was commissioned an ensign in the United States Navy. I would miss the pomp and circumstance of the spring ceremony. But that was okay. Mission Accomplished. Administered by the NROTC commanding officer, I took the oath of office in a small ceremony surrounded by my family, including my mother, who flew down from Michigan. My father, suffering from cancer, did not attend. It was similar to the oath I took in May 1961 when I first enlisted in the Navy: "I, David Francis, having been appointed an officer in the Navy of the United States, in the grade of ensign, do solemnly swear (or affirm) that I will support and defend the Constitution of the United States against all enemies, foreign or domestic, that I will bear true faith and allegiance to the same; that I take this obligation freely, without any mental reservation or purpose of evasion; and that I will well and faithfully discharge the duties of the office upon which I am about to enter; so help me God."

That day, I had that warm, fuzzy feeling. My mind and body were filled with an overwhelming sense of relief and pride. Soon my military pay soared to almost $10,000 per year, and I remarked to Linda that this was all the money we would ever need. I was wrong.

(L-R, Ensign Francis, Linda, Robin, Mildred and Jocelyn Francis)

Back to New England

The United States experienced a major snowstorm in January 1970. Never, ever move in the dead of winter if you can avoid it.

Graduating midyear meant that the Francis family of four needed to move across country in January. The mover hired by the Navy had a breakdown on the road and would arrive late. Our landlord become testy with us because he had a new renter coming in. It was a tense time.

After we finally got packed, Linda and I decided to make a stop in Maryland on our way east to Connecticut. We would visit Mike and Linda who were both students at the University of Maryland in College Park, Maryland. We drove off in our VW Beetle. Robin was three years old, and Jocelyn had just turned one year old and was still in diapers.

It was cold. Our inadequate heater did not warm us. We traveled with our coats on, and the windows fogged up frequently. I regretted throughout that trip choosing the Bug as our family automobile. But we arrived safely in Maryland and had a warm visit and a welcome opportunity to catch up with our friends from the *Bang* experience.

Mike was an art major and gave us the gift of one of his excellent nautical paintings. We've enjoyed it in our homes for many years.

It was cold and snowy in College Park, as it was in much of the United States then. As we were packing our car to leave for Connecticut, I failed to see that the patio door, once open, was now closed.

Trying to walk outside, I hit the glass door with Robin's Fisher Price See N Say. Several panes of thin glass fell down, cutting my hands and legs. I was bleeding, in pain, and embarrassed. The Lindas performed first aid but encouraged me to go to an emergency room in case I needed stiches. Mike drove me to nearby Andrews Air Force Base.

I will never forget the expression on the face of the Air Force physician's assistant who bandaged me. He must have been thinking, *What a stupid Navy ensign.*

We drove on to Connecticut. That is, Linda drove as I sat in the passenger seat, nursing my wounds. Our brief search for rental housing resulted in finding a very nice, three-level townhouse in Uncasville, an area between Norwich and New London. Once again, Linda pulled it off. Our goods arrived on schedule, and we set up our new Francis family residence.

We arrived in Connecticut a few months ahead of the next scheduled Naval Officer Submarine School, so I held temporary additional duty (TAD) orders to the Submarine Development Group Two (SubDevGru2), now Submarine Development Squadron Twelve. The Submarine Development Group developed advanced targeting tactics for US submarines.

We were happy to be back in historic New England again. And reunited with our dear friends, Raymond and Edna, who cared for us when we were young, inexperienced parents. Our townhouse neighbors on each side were Navy couples. Both men were ballistic missile submarine officers. The tide had changed. Both of their wives relied on Linda, the "veteran Navy wife," for advice and moral support when their husbands were at sea.

At the far end of our Uncasville fourplex lived the Gardners, an older couple who came to the United States from Yugoslavia. Mr. Gardner had competed in swimming at the 1936 Olympics in Berlin, Germany, and had fought with General Tito in the Yugoslav Resistance during WWII. It was always exciting and humbling to be with them. One evening, the Gardners hosted a dinner party. They served Turkish coffee. Linda and I enjoyed several cups of the delicious beverage. That evening we could not fall asleep. All night long, Linda and I talked and stared at the ceiling, feeling like amateur partygoers.

Experiencing and enjoying a "substantial" paycheck for the first time in my life and career, Linda and I happily traded our VW Beetle in for an Oldsmobile station wagon and purchased a large-screen color TV so Robin and Jocelyn could watch *Sesame Street* in color.

Linda loved the ride and the room of our first family station wagon, and we felt good about buying American this time, and helping our many relatives back home in Michigan who were still building "Detroit Iron."

What did I learn?

- ✓ Navy men and women take pride in their service and sacrifice. It is common and widespread in the seagoing service to feel a sense of personal esteem and value in choosing to wear the uniform and defend the homeland and its promise of "life, liberty, and the pursuit of happiness."
- ✓ The military oath of office is not to a person or a country. It is an oath of allegiance to the **Constitution of the United States.**
- ✓ "A good Navy is not a provocation to war. It is the surest guaranty of peace." —President Theodore Roosevelt.

- ✓ “I can imagine no more rewarding a career. And any man who may be asked in this century what he did to make his life worthwhile, I think can respond with a good deal of pride and satisfaction: ‘I served in the United States Navy.’” —John F. Kennedy.
- ✓ “When asked what I am most proud of, I stick out my chest, hold my head high, and state proudly, ‘I served in the United States Navy!”' —President John Kennedy.
- ✓ Never, ever move in the dead of winter if you can avoid it.

CHAPTER 15
OFFICER SUBMARINE SCHOOL

Image courtesy of https://www.netc.navy.mil/NSS

The Naval Submarine School (NSS) student manual says, "Through modernized training, the Naval Submarine School prepares officers and enlisted submariners with the resilience and knowledge necessary to be immediately successful on their submarines."[9] Naval Officer Submarine School in Groton was six months long. Despite my combination of academic and practical seagoing experience, I found it challenging and brought homework home every night. Sonar science was one of the most interesting and difficult courses. Back to math and physics again.

---

[9] Naval Submarine School, Naval Education and Training Command. https://www.netc.navy.mil/NSS/

Our class leader was a lieutenant and former swift boat captain in Vietnam. He was wounded in combat and had a metal plate in his skull. His combat service was rewarded with a set of orders to submarine school. Yet again, I felt fortunate to have avoided duty in Vietnam.

When it came to simulated submarine operations on trainers, my enlisted experience provided me a leg up. My long hours of standing watch at sea on the diving stand gave me hard-earned skill and confidence during simulations. On my first time acting as diving officer on the diving trainer, I quickly assessed the situation and brought the simulated boat into submerged trim within a few minutes. I savored the look of surprise on the trainer operator's face.

In class we discussed the loss of both *Thresher* and *Scorpion.* We listened to the audio tape from the submarine rescue vessel *Skylark.* It was heartbreaking to hear the often-muffled transmission from *Thresher's* commanding officer. We heard him report that *Thresher* was on its way up, then sliding back deep. Finally, we heard the sounds of the ship being crushed.

The fate of *Scorpion,* lost in May 1968, was still undetermined. One possibility is a tragic accidental explosion caused by a malfunctioning torpedo, known as "a hot run," torpedo explosion in *Scorpion's* torpedo room. The other explanation is more disturbing.

We learned that our Navy's top-secret sound surveillance system (SOSUS), a network of underwater acoustic sensors used to monitor and track both submarines and surface vessels, recorded an explosion, and that "a Soviet submarine was tracked leaving the area at a high rate of speed."

The US Navy Submarine Service is a small community. We sense collective heartache whenever we lose some of our own. I closely relate to the line from the movie *Star Wars*, "There's a disturbance in the force."

This time around, I had less fear of the required submarine escape procedure. My trip through the tower was easier and less life-threatening. I actually enjoyed the experience because I wore a Steinke Hood—an inflatable life jacket with a hood that completely enclosed my head, allowing me to breathe normally while ascending—no more "blow and go."

April 17, 1970

In class on April 17, I was approached by our class leader, who said, "Mr. Francis. I just received a call that your father has died."

In February that year, I had visited my dad when he was bedridden with cancer. I knew he was terminally ill, but it was still a shock.

Granted leave, Linda and I and the girls set out for Michigan. We tried to drive straight through, but there were just too many miles between Connecticut and Michigan for us. We arrived the day before his funeral. My dad would have turned fifty-six years old the next week, on April 24th. He was far too young. But was taken by cancer. I wore my dress blue uniform at the funeral out of respect for him. Military uniforms tend to add dignity to funerals, I believed. My mother reminded me often how proud my dad was of me and my military service. ("My son is a nuclear submarine sailor.")

That cold day in Michigan had a biting wind. We were unable to hold a graveside ceremony. We gathered indoors at Floral Gardens, a cemetery just outside of Bay City. Our young girls were so well behaved. For a long time, I embraced my fraternal grandmother, Myrtle Francis. This was the second time in her life that she buried one of her five children.

After our return to New London, I worked hard to catch up on my course work, but missing two weeks lowered my test scores and class standing. Competition between students for grades was rigorous and fierce and would determine the order of "boat selection" at graduation. Top students got the early choices of operating submarines that had open billets.

My previous Hemingway binge read in Uncasville didn't help any. This was long before cable and Netflix. After years of studying technical and scientific material, I decided to read for enjoyment. I read some Ernest Hemingway. In one weekend, I read *For Whom the Bell Tolls, The Sun Also Rises,* and stayed awake all night because I could not put down *The Old Man and the Sea*. Perhaps I should have been reading about sonar.

The Country Club

That summer, the entire Francis family enjoyed the recreation and dining that the Submarine Base Officer's Club offered. It is similar to a country club, one of the many perks of commissioned service. One evening Linda and I were at the O Club, enjoying a live band

and a banquet. As the band started playing *Moonlight Serenade*, a Big Band number I knew, I turned to Linda and asked, "Would you like to dance with a dashing young naval officer?"

Looking around the room, she responded, "Yes. Where is he?"

Our girls spent many hours at the base swimming pool. In August, my mother flew out for my graduation. She loved to play golf, so I took her to the Submarine Base Golf Course (just like a country club). We both teed off on hole three, a short drive. But then I could not find my ball. As I was searching in the weeds, my mother putted in.

Suddenly I heard her yelling as she found my ball in the cup.

"You hit a hole in one! You hit a hole in one!"

"Mother. Please be quiet. I don't want to buy all the senior officers a drink at the bar. I'm just a lowly ensign."

We quietly finished our game then returned to Uncasville.

Graduation day arrived. The ceremony prescribed that we wear our service dress white uniform ("choker" whites) with sword. I had the right uniform, but I owned no sword. Fortunately, I was able to borrow a sword from my Naval Academy neighbor.

My class standing was in the lower 50 percent. I had managed to underachieve. Consequently, at the graduation luncheon and "boat selection," I would be in a lower pool and get what remained on the list of submarines after the top students chose. When my name was called, I selected the USS *George Washington*, the US Navy's first Polaris missile submarine, named in honor of our first president: "First in war, first in peace, and first in the hearts of his countrymen."

What did I learn?

- ✓ Santiago, the Cuban fisherman in Hemingway's *The Old Man and the Sea* lamented that he "went out too far." I lamented that I rested on my oars. I didn't do my best for me, my family and my career. I didn't work hard enough in Submarine School.
- ✓ "To give anything less than your best, is to sacrifice the gift." — Steve Prefontaine.
- ✓ Steve Roland "Pre" Prefontaine was an American long-distance runner who competed in the 1972 Summer Olympics. While running for the Oregon Track Club, Prefontaine set American records at every distance from 2,000 to 10,000 meters. Pre was a hero to our son, Jon, a college runner.

CHAPTER 16
POLARIS

Polaris: Oil on canvas by James E. Mitchell, 1970, courtesy NHHC #70-731-B

Our Navy named a missile, a class of submarines, and a massive weapons development program after the North Star, Polaris. President Eisenhower in his January 1961 farewell address, in the tradition of George Washington, warned the nation against the accumulation of power by a military-industrial complex. Eisenhower was legitimately concerned that long-term Cold War mobilization would harm the US economy.

However, in his duty as President of the United States, Commander-in-Chief, and his strong desire and commitment to protect US national security, President Eisenhower had approved, during his administration, one of the most expensive, complex, successful, and historic, military-academic-industrial projects.

His military development effort that catapulted the United States ahead of the Soviet Union in the arms race was called the Polaris Development Program.

The purpose of the program was to place ballistic missiles at sea aboard submarines. Hidden at sea, our missiles would be able to survive a potential Soviet first strike. Land-based bombers and missiles (even in silos) were vulnerable. Undetectable submarine launched intercontinental ballistic missiles (ICBMs) at sea would provide the US with greater national security.

The Polaris Development Program, begun in 1955, is legendary in naval history for its exceptional project management and contribution to submarine capabilities. Rear Admiral William "Red" Raborn led and managed multiple US industrial, academic, and government agencies including MIT Instrumentation Laboratory, Johns Hopkins Applied Physics Laboratory, Lawrence Livermore National Laboratory, Lockheed Martin, Honeywell, General Electric, and Electric Boat.

Significant new technologies emerged including Ships Inertial Navigation System (SINS) and a satellite navigation system (NAVSAT) for more accurate ship's positioning. A computer small enough to fit through the submarine hatch and lighter nuclear warheads were also developed.

Enhanced missile guidance and ships positioning systems were vital for missile accuracy to target. To house sixteen Polaris Missiles, the original *Scorpion* fast-attack submarine, under construction at Electric Boat in Groton, was cut in half, and a missile compartment was added.

Our older diesel electric submarines like *Bang* were designated SS. Nuclear submarines like *Seawolf* are SSNs. N for nuclear. Our ballistic missile submarines added a B for ballistic—USS *George Washington*, SSBN 598.

The lead ship of the class, *George Washington,* with a crew of about 140, was 381 feet long and displaced 6,000 tons. Since she was a modified Skipjack class fast attack, she had six 21-inch torpedo tubes forward and sixteen Polaris missile tubes in the missile compartment. The first Polaris A-1 had a range of 1,400 nautical miles. Refinement of Polaris from A-1 to A-2 and A-3 improved range, accuracy, and payload (the number and megatonnage of nuclear warheads).

On July 1960 *George Washington* successfully launched its first Polaris missile and made her first nuclear deterrent patrol in November 1960. Four more *George Washington* class SSBNs were built between 1959 and 1961: USS *Patrick Henry*, USS *Theodore Roosevelt*, USS *Robert E. Lee*, and USS *Abraham Lincoln.*

During the Cuban Missile Crisis in October 1962, Eisenhower's successor, President Kennedy, was a beneficiary of the Polaris Development Program (from Chapter Four: The Brink of War). Nuclear-powered vessels can remain at sea for years. They have almost unlimited fuel capacity and range. However, the crew cannot stay at sea as long as the ship can. Consequently the SSBNs established the two-crew concept, Blue and Gold, to significantly increase the number of days the submarine could be at sea, on patrol, covering its target package.

Crews can be deployed for up to 105 days and at home for 95 days, with a three-day turnover on each end of the deployed period. Crews are flown from home bases to and from the forward deployment bases. After taking over the boat, the new crew performs a thirty-day refit assisted by the tender, followed by a 62 to 70-day deterrent patrol. In the early years, the home bases for Atlantic Fleet crews were Groton, Connecticut, and Charleston, South Carolina. Pacific Fleet crews were based at Pearl Harbor, Hawaii, and then Guam. To get closer to the assigned targets, with our short-range Polaris missiles, forward bases were negotiated at Holy Lock in Scotland and Rota, Spain on the Bay of Cadiz. Submarine tenders were stationed at each SSBN port.

In August 1970, Ensign David Francis, was assigned to the USS *George Washington*.

What did I learn?

- ✓ I am grateful to US taxpayers who invested heavily into my education and career.
- ✓ Military service is a common path to upward mobility.
- ✓ The Polaris Development Program brought the United States significant safety and security.

CHAPTER 17
DAM NECK ON THE ATLANTIC

My new ship, the first submarine I served aboard as a commissioned officer, the USS *George Washington*, was sitting in dry dock at the Charleston Naval Shipyard in Charleston, South Carolina. Her homeport was New London, Connecticut, but she was undergoing an overhaul and reactor refueling in Charleston.

My orders included "temporary additional duty" at the Polaris Weapons School in Dam Neck, Virginia, near Virginia Beach, where I received training to serve as a Polaris submarine weapon's officer.

My commissioned service was primarily the result of shortages. The Naval Enlisted Scientific Education Program (NESEP), that provided my college education and commission, was created because there was a shortage of naval officers in the 1960s, especially technically trained officers and carrier pilots.

My assignment to a nuclear submarine was caused by a shortage of nuclear trained officers (Nukes). The US Navy's nuclear fleet was expanding, and there were not enough Nukes to fill the billets and operate the Navy's nuclear reactors. Consequently, officers were assigned to submarines who could serve forward of the reactor room. Weapons, Communications, Supply, and Navigation officers, for example, did not require nuclear training and qualification.

Since the Francis family would eventually end up living in Charleston, South Carolina, we decided to move out of Uncasville, Connecticut, put most of our household goods in storage, and spend some beach time together in Virginia Beach. We rented an efficiency apartment a few blocks from the Atlantic Ocean.

During the day in Virginia, I went to weapons school, and Linda and the girls moved back and forth, to and from the room, the pool, and the beach. It was an enjoyable lifestyle at first, but the adventure grew old as the weather cooled in the fall. The togetherness in a two-room efficiency was a challenge. But Linda managed it with grace. She was pregnant with our third child; and Jocelyn, less than two years old, was still in diapers. I was in awe of how Linda handled it all so well. There were some welcome breaks visiting with my Aunt Thelma, Uncle Ned, and their kids, our cousins Susan and Johnny in Norfolk. Ned was now retired Navy.

Polaris Weapons School was ten weeks long. We studied missile components and the launcher and fire control systems.

After my graduation in November, we headed down to Charleston to look once again for a new home—our sixth move in six years.

What did I learn?

✓ I again learned that I married an amazing woman who managed the challenges of the military life with skill and grace.

CHAPTER 18
CHARLESTON

In November 1970, the Francis family moved to Charleston, South Carolina. Charleston was rife with military installations, including Charleston Naval Base, Charleston Air Force Base, and the Charleston Naval Shipyard. Congressman L. Mendel Rivers was enjoying a long tenure as the Democratic Representative from Charleston's 1st Congressional District (16 terms). Rivers was the longtime chairman of the House Armed Services Committee and a strong supporter of the US military who managed to bring much of it to South Carolina.

We were told in New London that *George Washington* would return to Connecticut at the completion of her overhaul. After a short search using a realtor, we purchased our very first house. Our three-bedroom, one-level brick rambler reminded us of our Louisville rental. The house was a bank foreclosure in the Pepper Hill subdivision of North Charleston near the Charleston Air Force Base. It needed some cleaning, but the price was very reasonable. Linda, five months pregnant, with two children under four years old did her "magic act" again to establish our new nest.

I reported aboard my new ship. That is, since *Washington* was sitting in the drydock, I arrived at the shipyard building that housed the ship's offices.

My new chain of command was Don, the commanding officer; Keith, our executive officer; Bill, the weapons officer; and Jim, our assistant weapons officer (AWEPS). I would report directly to Jim.

My duties were to serve as deck officer (first lieutenant) and torpedo officer, help get the ship through the yard period, and begin my submarine weapons and officer submarine qualifications (i.e., earn my gold dolphins). My additional duty was barracks officer to oversee good order, discipline, and cleanliness in the barracks where several *George Washington* sailors lived.

Similar to my experience on *Seawolf,* I assembled a large Submarine Qualifications Notebook with all of the knowledge and signoff requirements for ship's systems, operations, procedures, weapons, and casualty and damage control. I also took charge of monitoring shipyard repairs of deck areas, superstructure (outside the pressure hull), and the torpedo room. All that would keep me busy.

About two-dozen seaman and torpedomen populated my two divisions. Some had college degrees. That was not surprising since they enlisted in the Navy, volunteered for submarine duty, and the war in Vietnam was still grinding on.

Shortly after reporting aboard *George Washington,* I was promoted to lieutenant junior grade. Jocelyn gave us a small blessing by being potty-trained before her second birthday. We wouldn't have two children in diapers.

Linda and I were fortunate to find a new supportive social group. Jim, my boss, was a NESEP officer, a graduate from MIT, and one of the most-capable naval officers I ever served with. He possessed a rare combination of kindness, brilliance, and common sense. Jim and his wife Diane became our good friends. Keith, the executive officer, and his wife Bonnie were also Episcopalians, and we attended church together.

We were part of a "wardroom" for the first time, members of a small group of ship's officers and wives (and children). It was a busy and challenging time for me, with many ship repair and upkeep duties to perform, work on my submarine and weapons qualifications, and learning how to be a competent division officer.

One day, I was summoned by the shipyard superintendent to inspect and approve new piping welds on the boat. Never in my life had I been trained or required to inspect pipe welding. Since a ship's officer needed to sign off on each job, I sought his advice.

He said, "Inspect the x-rays taken and look for any areas that don't look smooth or solid. Look for voids or bubbles."

Remembering that a suspected cause of the *Thresher* sinking may have been a bad piping weld, I became very good at that task. My inspection skill would be put to the test when our ship went back to sea.

In January I received a letter reminding me that we needed to have two qualified scuba divers as part of the crew before we left the shipyard. Ship's divers were needed to go into the water to inspect the hull or retrieve people or objects. I asked for volunteers and selected two eager, young seamen. Off they went to New London for a two-week Navy Scuba Diving School.

After a week, one of them returned to Charleston.

"Why are you back so soon?" I asked.

"Mr. Francis. I need to speak with you."

We went into my office, and he began to talk. The young man said he was standing by the pool, when it started on fire, and his scuba instructor turned into the devil. I was speechless.

After a long silence, I asked, "Why do you think that happened?"

"It was an LSD flashback."

"Where did you get the LSD?"

He told me he got it from a woman he met in New York's Time Square. OMG!

Feeling disappointed and angry, I said, "You know, not only will you be kicked off this submarine, but probably out of the Navy."

He agreed and said he was sorry. The incident turned into a full day of medical and legal meetings and administrative paperwork for me. And I needed to quickly find a new drug-free volunteer. *Well. I wanted to be a division officer.*

My required annual physical examination detected another inguinal hernia. As directed, I checked into the Charleston Naval Hospital for the hernia repair. In pain for several weeks and having difficultly climbing stairs, I stayed off the ship and avoided laughing, because it hurt to laugh.

After recovering from my surgery, to prepare for the birth of our third child, I sanded and varnished the floor and painted the walls of our third bedroom that would be the nursery. I installed a partial wooden fence and planted more grass in the backyard to make a better playground for the girls. My neighbor advised me to plant centipede grass. Centipede grows quickly, makes a thick green turf, doesn't require a lot of maintenance and thrives in South Carolina.

In short time we had a great backyard. For the first time in my life, I felt like I had a green thumb.

Melissa Sue Francis

Linda went into labor on March 17, St. Patrick's Day, so we went to the hospital. St. Francis, the civilian hospital in Charleston, unfortunately did not allow fathers to be present during birth. I returned home to be with Robin and Jocelyn. Linda did not stay in labor for long. Within a few hours, we were parents of a third daughter. By the grace of God, we were blessed with three beautiful, healthy daughters. Three children in five years! Our first daughter was born and baptized in Connecticut, our second in Kentucky, and our third in South Carolina. A real Navy family.

My mother flew down from Michigan to help with the girls for a few weeks. Linda was stretched but did not break. She had three daughters at home all day until the fall, when five-year-old Robin started kindergarten.

Since South Carolina did not have public kindergarten, Linda enrolled her in a Baptist Church School. Each morning Robin took White Bus #1 to kindergarten.

What did I learn?

- ✓ Be careful what you wish for. Serving as a commissioned officer is challenging work.
- ✓ Linda soon took an interest in the history and beauty of Charleston and the surrounding Lowcountry. Similar to our time in New England, she encouraged me to share in her new discoveries.
- ✓ Take the time and effort to learn about and appreciate different regions, cultures and environments.

CHAPTER 19
WASHINGTON

**USS *George Washington* SSBN-598,
courtesy NHHC, catalog #L45-108.08.01.**

Serving on a nuclear submarine, carrying nuclear weapons, there are lots of inspections. After completing one inspection, you begin preparing for the next one. My first formal inspection as a submarine weapons officer did not involve naval or nuclear ordnance. As first lieutenant/deck officer, I was responsible for submarine salvage and escape systems. These features are necessary to facilitate a crew escape from a sunken or disabled submarine, which can threaten the safety and lives of the crew from flooding, lack of oxygen, and a toxic atmosphere. The most important systems were the forward and after escape trunks, escape appliances (like Steinke hoods), and the deck connections for the deep submergence rescue vessel (DSRV).

The DSRV is a manned, mini submersible that can go down to a sunken submarine and attach itself over the top of the escape trunk hatch. Rescued submarine sailors can then enter the DSRV for the ride to the surface.

Since I was insanely worried about failing my first inspection as a ship's officer, I devoured the Naval Sea Systems Command technical manuals for guidance and requirements. I worked, sometimes annoyingly close, to the shipyard superintendent to make sure that all shipyard work was complete, and the system tests were done satisfactorily. Then I sent in my seaman to clean, prepare, and paint all required spaces and areas. I stayed on top of it and passed the inspection. That really felt good.

Blue and Gold

United States Navy colors are blue and gold. So it was natural for the submarine force to use blue and gold to identify each crew in the two-crew ballistic missile submarine concept. As *George Washington* approached the end of her shipyard refit, more personnel came aboard and were assigned to fill the billets for each 140-man blue crew and gold crew. Back then, only males served aboard nuclear submarines (not like the coed crew of Starship Enterprise).

I was assigned as Gold Crew assistant weapons officer. Keith was our executive officer, and Jim, our weapons officer. When our Gold Crew commanding officer arrived, Lieutenant Commander

David (Dave) Fields, at thirty-five years old, was the youngest nuclear submarine commanding officer in the fleet. He was nearly the same physical size as me, very personable and high energy.

Word came down. (That's where word comes from) that *George Washington* would not be returning to New London as her home port. She would stay in Charleston, meaning that's where the crews and families would be stationed.

Many, including Linda and I, were deeply disappointed at the news. We would not be returning to New England as hoped. We really wanted to return to Connecticut. Linda and I loved New England history, landscape, and culture. This was the third time in my naval career where a set of orders made me feel both sad and mad.

George Washington would now be physically located at the Naval Base in Rota, Spain, and conduct ship's refits there prior to going to sea for deterrent patrols. Rota was one of our advanced bases in Europe that put US Polaris missiles "closer in" and reduced submarine transit time to get on station and cover our assigned target package.

The Blue Crew, the senior crew, took *Washington* to sea first. We took over the ship in the fall of 1971 to conduct our post-overhaul shakedown operations and our nuclear weapons acceptance inspection (NWAI). The NWAI is a rigorous inspection to certify the ship and crew for nuclear weapons storage, handling, and deployment.

We got underway and headed south for Florida. It had been over four years since I was last at sea. After we arrived at Cape Canaveral we were met by the NWAI inspection party from Navy Special Projects Office in Washington DC. Special Projects developed the Polaris Missile System and is the nuclear weapons equivalent to the US Naval Reactors Office (Admiral Rickover's shop).

The inspection team was composed of several experienced submarine nuclear weapons officers. They were sharp guys. This inspection proved to be the most stressful, fast-paced, two weeks in my life. We began early each day, went to sea, and went through a series of simulated missile-launch procedures and casualty drills.

My battle station was at the launcher control panel located in the missile compartment, outboard and between missile tubes #3 and #5. The missile compartment had eight odd-numbered missile tubes to starboard and eight even-numbered tubes located on the portside.

As the inspectors would insert several simulated casualties into the system, we were required to assess and correct the problems then continue with the launch sequence. I was fortunate to have a highly capable chief petty officer at the panel and frequently asked, "What's the problem, Chief?"

After two grueling weeks of training and inspection, we loaded an actual Polaris missile into missile tube #5, with telemetry in its warhead.

Out at sea on the Atlantic Test Range we were accompanied by The USS *Compass Island*. Her mission was to track the flight of our missile. By then, we "*Washington* Weaponeers" were near flawless in our fire control and launcher sequence procedures. Stationed next to missile tube #5, I heard the unmistakable noises of the rocket rising out of the tube. It flew accurately and landed in the ocean right where it was supposed to.

We passed our inspection and were certified to carry, handle, store, and launch Polaris nuclear missiles. Captain Fields graciously presented to our weapons team a color photograph of our Polaris missile breaking the surface of the ocean. Our sense of relief and elation was overwhelming.

Tight Spaces

We returned to Charleston to tackle our next challenge, an inspection to certify George Washington Gold Crew to handle, store, and fire torpedoes. Washington was an aging boat with frequent mechanical problems.

We were alongside the submarine tender in the Cooper River and the tender had repaired one of *Washington's* malfunctioning torpedo outer-door mechanisms. The job required an officer inspection and signoff. As torpedo officer, I got the assignment.

If you have claustrophobia, you could not serve aboard a submarine. Many spaces are tight. In Submarine School you are placed in small, pressurized places to confirm that you can tolerate it. But there was a time when I felt a tinge of claustrophobia.

It was uncomfortable and scary to be lowered headfirst through an opening in the superstructure, below water level, down a dark, slimy, tight space to the bottom of the ship. With a flashlight in hand and headphones, I witnessed the operation of the mechanism and reported proper functioning. Then I was pulled back up, feet first, safely to daylight.

You might recall from watching old submarine movies that our World War II diesel electric fleet submarines sent compressed air directly into the torpedo tube to expel the torpedo. Unfortunately, released bubbles would then rise to the surface, exposing the position of the attacking boat. The submarine force solved that problem by designing a system that pushed the torpedo out of the tube with seawater rather than air.

The firing air was released into the torpedo room through a muffler to dampen the sound. It was a sophisticated system with many moving parts. That brings me to my most embarrassing day as a submarine officer. We were at sea and the squadron commander was on board.

Word was passed on the ship's general announcement circuit (The 1MC), "The ship will be firing water slugs."

The Check List

My team and I assembled in the Torpedo Room, eager to demonstrate our skills and pass our inspection. Several observers stepped into the torpedo room, including our commanding officer and the squadron commander.

Following the required checklist, I directed the firing. The firing handle was pulled. Air hissed violently through the system. Bang! The piston drove forward. Then swish, the sound of rushing water in the torpedo tube.

Boom! Success? No!

Firing air was loudly released, and the torpedo room was instantly filled with small white flakes as if we were in a snowstorm.

Hundreds of white flakes landed on everything and everyone including the captain and the commodore.

Stunned, I was met by an angry look from the CO. He said, "Dave. Find out what happened here." He spun around and departed. My senior torpedoman, was the first to assess the confusing situation.

Following his hunch, he opened a deck plate and went down to inspect. In the Torpedo Room bilge, he found an exploded case of Nabisco Vanilla Wafers.

Further investigation led to the knowledge that a seaman, ignorant of the function of the muffler, hid the entire case of cookies there, intending to hoard them. It took several days and many seaman and torpedomen to clean up the mess.

Our torpedo firing checklist contained all required steps except one: *check inside the muffler for cookies.*

Soon full-color, hand-drawn depictions of the event were posted throughout the ship, showing a torpedo room filled with white flakes and an officer (me) announcing, "The ship will now be firing *cookie* slugs." *Washington's* political cartoonist remained anonymous.

After several more days of observation and questioning by weapons' inspectors and completion of torpedo makeready procedures, we were certified to carry torpedoes, including a few nuclear tipped "fish." The torpedoes with nuclear warheads were actually the easiest to maintain. My divisions had now passed all of our weapons' inspections. This accomplishment was reflected in my first officer fitness report from the Gold Crew captain.

With a complete lack of modesty, let me provide you with an excerpt: "Commander Submarine Squadron Sixteen commended the ship for . . . and the excellent spirit displayed by the Weapons Department during the examination. LTJG FRANCIS directly contributed to the positive attitude displayed by his men."

Officer fitness reports (job performance evaluations) are important for officer promotion, retention, and assignments. They are like Olympic medals. We strive for the gold.

Loaded for War

We were now certified to carry torpedoes, so we tied up again next to the submarine tender in Charleston and prepared the forward deck for the load. One by one the crane swung over a torpedo in a sling. We received it, placed it on the greased skid, and lowered it tail first, the old-fashioned way, with ropes and pulleys into the torpedo room.

Torpedomen in the room moved it to a skid and locked it in place. *Washington* carried a mix of torpedoes: a few with nuclear warheads, several MK 37 battery-powered electric motor driven, and some Mark 14s. The Mark 14 is the torpedo that "won the war in the Pacific," sinking thousands of tons of Japanese ships. It's an old, long, heavy, steam-turbine-driven weapon, fueled by alcohol.

As torpedo officer, I monitored the evolution topside with frequent trips to the torpedo room to check on safety and placement. We worked from morning until dark. Finally I reviewed and signed the torpedo custody forms. We now owned a room full of real torpedoes.

While I was initialing the custody forms with my initials, DLF, my leading torpedoman said, "Mr. Francis it's good thing your parents didn't name you Edward."

I was exhausted and asked, "Why is that?"

"Because your initials would be ELF." Yup. Another "short" joke.

Unfortunately, the Gold Crew engineers did not fare as well. The Nukes failed their Operational Reactor Safeguard Examination (ORSE). The ORSE is conducted by the Office of Naval Reactors' personnel who inspect for crew readiness and training and the operational safety of the nuclear reactor and engineering plant.

Consequently, members of the US Naval Reactors Office rode *Washington* for several weeks directing a "get well" program to correct deficiencies and to conduct a reexamination. It was a tense time for all onboard. After several weeks, the engineers passed their reinspection.

Our next stop on the Cooper River was the Charleston Naval Weapons Station established in 1941. In 1960 the Polaris Missile Facility Atlantic (POMFLANT) was added to the Weapons Station to store and provide Polaris missiles. Our sixteen Polaris missiles did not swing over on slings like our torpedoes. They were encased in containers. We opened the missile tube outer door and secured the container on the top of the tube then slowly lowered the missile down with an electric wrench. It was an all-day, all weapons department evolution.

At the end of the day, we took ownership of sixteen intercontinental ballistic missiles with thermonuclear warheads. Finally *George Washington* had her full weapons load.

We were now in the strategic nuclear weapons business, a participating member in the mutually assured destruction strategy and custodians of our nation's primary peace-keeping mission.

The Blue Crew took the ship across the Atlantic Ocean for her first post overhaul strategic missile patrol. Following her patrol, *Washington* would be tied up at the Naval Base at Rota, Spain.

What did I learn?

- ✓ I learned that to be successful as a naval officer, it was vital to respect, empower, and reward the sailors who did the work and "carried the load."
- ✓ I was fortunate to be supported and surrounded by US submarine sailors, some of the best-trained and most-talented sailors under the waves.
- ✓ Positive affirmation, even encouraging self-talk, along with praise, verbal and written, promote human motivation and diligent effort.
- ✓ Praise in public; discipline in private.
- ✓ The most important words that a naval officer must learn are: "Hey, Chief, I have a problem and I need your help."

CHAPTER 20

THE MAGIC KINGDOM

"Where your fantasy becomes a reality"

What do you do when you've passed all of your Weapons Department inspections? You go to Disney World. While I was single-mindedly focused on my career and shipboard duties, Linda was planning our next adventure. Gold Crew was now the off crew. We were ashore in a training status in Charleston, until we would take a crew flight to Rota, Spain, to relieve the Blue Crew.

Keith and Bonnie, brave and gracious, agreed to watch Melissa, our "baby" daughter. The four of us drove in our station wagon south to the Magic Kingdom. Five-year-old Robin walked. I pushed three-year-old Jocelyn in a stroller. They may have been too young to appreciate the experience, but we were going to provide it anyhow.

Linda and I were reunited with Disney's "It's A Small World" exhibit that we enjoyed at the World's Fair in 1965. The boat ride through animated singing characters was a favorite of the girls.

Robin was frightened by the queen/witch in the Snow White exhibit, and we did not return to that one. Linda and the girls rode the Teacups many times. Because I suffer from motion sickness, I just watched. The experience for me was culture shock. I went from a high stress, "failure is not an option" business to Fantasyland.

Reluctantly we returned to realism in Charleston, where I reentered the path toward multiple qualifications I needed to achieve: weapons duty officer, diving officer, officer of the deck, and qualified submarine officer (earning my Gold Dolphins).

Weapons duty officer was first. Since there were only two of us, me and Jim, the weapons officer, I needed to get signed off so Jim could get a day off from standing watch. In port, there must be at least one qualified weapons duty officer onboard to ensure nuclear weapons safety and security. Off crew gave me opportunities to study the manuals, pass the tests, and get signatures on my many qualification cards.

Diving officer, officer of the deck, and "officer submarine qualified" required considerable time and effort at sea since they are ships' operations, and casualty procedures intensive. Meanwhile I could get lots of reading done. I spent time on the ship's trainers/simulators available to us on base and was convinced that my enlisted submarine service helped me to avoid feeling overwhelmed.

Firearms

As assistant weapons officer, I was responsible for the training, care, and safety of all firearms on board. All Navy ships carry firearms for protection of the ship and the crew against outside attack. One of the most common shipboard drills is "repel boarders" where the crew removes firearms out of the gun lockers and take their repel boarder stations to defend the ship.

As a young seaman, I stood topside watch, many days and nights, for many hours, with a Colt .45 service revolver on my hip. I could not escape the irony of a teenager with a pistol defending a multimillion-dollar nuclear submarine.

Off crew we went to the firing range for training and certification. Fortunately we no longer carried the M1 rifle I fired in boot camp. We trained with the M-14, the M-16, the Colt .45 caliber service revolver, and even the Thompson submachine gun. The Thompson, invented during World War I, became famous during 1930's Prohibition in America as the gun of choice by criminals and bank robbers. On board I was responsible for the inventory of all firearms and kept a .45 Colt in my safe in case of mutiny. Of course, it was highly unlikely that there would ever be a mutiny on a US Navy nuclear submarine. And there wasn't.

West of the Ashley

With the certain knowledge that we would remain in Charleston for a few more years, Linda began looking for a larger house in a quieter neighborhood. We wanted to be closer to our friends, the Charleston Naval Base, and get out of the landing pattern for the Air Force Lockheed C-5A Galaxy transport planes that descended into nearby Charleston Air Force Base. Many times, I could see the faces of the pilots as they flew the massive aircraft over our house.

Linda found a beautiful three-level home, west of the Ashley River. For the first time, we used my Veterans Administration Home Loan. The house was move in ready. We were unable to sell our North Charleston home in time, so we rented it out to a Navy chief and his family. We went from renters ourselves to owning two houses in two short years. Linda and I agreed that we should have purchased our first home much earlier in our marriage. In December 1972 we marked our ten-year wedding anniversary in our new west-of-the-Ashley home.

What did I learn?

✓ Spending rest and recreational (R & R) time with your spouse and children is important for everyone's mental health, feelings of recognition, and rewards for their stress and strain.

CHAPTER 21
FIRST PATROL

The SSBN Deterrent Patrol Insignia is a uniform breast pin worn by officers and men of the United States Navy's submarine service who have completed strategic deterrent patrols in nuclear ballistic missile submarines. It is the successor to the Submarine Combat Patrol Insignia awarded for submarine patrols during World War II.

The year 1972 marked my eleventh in naval service. During those years, I served aboard three submarines, graduated from college and several service schools, earned my silver dolphins, experienced two shipyard overhaul periods, participated in NATO exercises, sea trials, torpedo and missile loads and launches, multiple nuclear weapons inspections and tests, and advanced through seven military pay grades—from seaman recruit to lieutenant (junior grade).

I rose from ship's mess cook to ship's officer. It was an eventful journey. Next in my naval adventure would be my first 104-day deterrent patrol aboard an SSBN. It would be my longest time submerged and my longest time away from my family.

With a deep sense of sadness and dread, peppered with a bit of excitement, I hugged and kissed Linda and our girls goodbye at the Charleston airport. We boarded the crew flight to Spain. It was a seven-hour flight. By the time we landed in Rota, I had been awake for sixteen hours. (I have never succeeded in sleeping on airplanes.)

Gold Crew rode a bus to the Rota Naval Station then boarded a tugboat to go out to sea to meet *George Washington* waiting for us in the Bay of Cadiz. On board we began our turnover with the Blue Crew. It was busy and crowded, with over 280 people on a submarine built for 140. Diligently I followed my checklists, interviewed my counterparts, and conducted inventories and sign offs.

After twenty-four hours of constant activity, my brain was telling me that it needed sleep. But I pressed on. After thirty hours awake, I was able to complete my tasks and gratefully crawled into a bunk in the torpedo room in a berthing space called the "hanging garden:" a cluster of ten elevated bunks.

After our change of command from the Blue Crew, we set to work on planning and conducting our thirty-day refit period to get the ship ready for our first patrol. *Washington* at fourteen years old was an old submarine with lots of miles and months at sea. She needed frequent repairs, maintenance, and upkeep, so we created many work orders for *Simon Lake,* our Rota submarine tender.

During this refit period I completed my weapons duty officer qualifications and could now stand in port watch, alone, and allow Jim, my weapons officer, to take a day off. However, it was port and starboard (every other day) duty since there were only two of us.

A handful of times, I got off the ship and made it to the Rota Officer's Club. Mostly it was long, sixteen-hour workdays with a long task list. Aging ships, I learned, show "signs of wear," have multiple mechanical problems, and require extensive maintenance attention.

One brief visit to the Officer's Club was particularly memorable. Sitting at the bar, minding my own business, a young woman came up and asked me to dance with her. I obliged. The DJ was playing *Sweet Caroline* by Neil Diamond. At the end of the song I thanked her, and asked what brought her to Spain.

She replied, "My dad is the admiral on base."

I thought it was best for me to go back to the ship. I said goodnight to the young lady and left.

War Patrol

After a rigorous refit period, we got underway. Going out to sea never felt so good. "You have been deloused." On my first underway on a deterrent patrol I was introduced to the delousing operation. As you know, a louse is a blood-sucking insect. One of our own very quiet, fast-attack submarines trailed us, searching for the presence of lice, meaning a Soviet submarine, which may be trying follow us to our patrol area.

United States Navy missile submarines are the primary deterrent to nuclear war because of our stealth. We can remain hidden from our enemies. If *George Washington* fell prey to detection and tracking by a Soviet submarine, we would be vulnerable and exposed. The delousing procedure ensured that we were "lice free."

We would pass through the Straights of Gibraltar and begin our two-month patrol in the Mediterranean Sea. Our mission was to remain undetected and cover our target package with our sixteen Polaris missiles. The commanding officer informed me that we were making a "war patrol" and ordered me to make ready and tube load six war shot torpedoes.

My torpedomen and I spent two full days in the "room," conducting makeready procedures. The Mark 14s were the most time-consuming since we needed to add fuel and arm the warheads. Battery-powered Mark 37s were easier to make ready. Since *Washington* was a 1950s vintage fast-attack, tube loading was done by hand with ropes and pulleys.

We attached the A cables (data transmission cables) and placed the wire canisters behind each wire-guided torpedo. Finally, we placed a sign on each torpedo tube door that read, "War Shot Loaded." In the unlikely event of hostile action, we would be ready to rapidly return torpedo fire at our attacker(s).

During this time, I gained more appreciation and pride for "my" submarine torpedomen. They possessed intelligence, physical strength, and *esprit de corp*.

Predictably, hand-drawn color drawings appeared, created by Washington's anonymous political cartoonist, giving *Washington's* commanding officer the nickname of "Captain War" with captions of "George Makes Her First War Patrol."

Similarly, I was required to close out all sixteen Polaris missile warheads by crawling inside each warhead, carrying a schematic diagram to ensure that the thermonuclear warhead wiring and components had not been tampered with. After I exited, I sealed each door shut, then officially certified that I observed no evidence of sabotage.

Inside a few warheads I found some graffiti, compliments of the weapons' contractor. One read: "To Russia with Love" and another, parroting the Hallmark greeting card, wrote, "When you care enough to send the very best."

Like many nuclear weaponeers, I maintained my sanity by not dwelling on the millions of casualties and the immense human suffering we could inflict. But instead, I held to the firm and comforting belief that our mission was to deter a nuclear war and to keep the peace.

Mediterranean Cruise

The navigator had laid out an interesting route for our sixty-two-day patrol in the Mediterranean. We would pass along the North African coast, continue to the east, circle around Cyprus, then work our way through the Aegean Sea. Take westerly courses around Greece then "sneak" by Sicily. Steam WNW offshore of Italy. Head west past Southern France then go south off the Spanish coast. It was like a Mediterranean cruise. However, we would not be lounging topside in deck chairs, enjoying the sun. We remained submerged and hidden.

We ran below the surface of the ocean waves, level (zero bubble) at three knots. It was a smooth and comfortable ride. The Mediterranean Sea was mostly calm during that time of year. Since I was prone to motion sickness, I frequently thought, *Now this is the way to go to sea.*

After my weapons work was done, I began standing daily watches as diving officer and officer of the deck under instruction. My goal was to qualify at both stations before end of patrol. Captain Fields and several senior-qualified chiefs and officers were eager to help. More watch standers meant fewer hours on watch for everyone.

As a result of my months as a submarine seaman standing helm and diving planes watches on *Seawolf,* and the generous mentoring from many chiefs of the watch, I soon qualified as diving officer.

Luckily, in my approaches to periscope depth, we encountered calm seas and gentle swells. I did not have to work hard to keep the boat in trim and at our assigned depth.

As a qualified watch stander, I now felt like a contributing member of the crew. However, that doubled my time in the Control Room, standing a four-hour watch as diving officer and then another four-hour watch as officer of the deck under instruction (OODUI).

As officer of the deck (UI), I often took the boat to periscope depth. The view of the outside world relieved some of my sensory deprivation, and as a bonus I was able to see from a safe distance the coast of North Africa, Greece, Italy, France, and Spain. The beautiful azure waters of the Aegean were spectacular.

Wild Game Night

Washington's resourceful supply officer (SO) discovered that Captain Fields enjoyed eating wild game. So he ordered and stored rabbit and duck meat for our patrol. It was available in the Navy supply system. The captain declared Thursday evening as Wild Game Night in the wardroom. Dutifully, all officers not on watch attended the first one. Not a fan of wild game, I poked at my duck, ate little, and went away hungry.

At the next Wild Game Night, fewer officers attended. The third week, the captain dined alone. Getting the message, he informed his officers and the cooks that Wild Game Night would be discontinued. We returned to more traditional submarine cooking.

With watch standing, division work for three divisions, submarine qualifications study, and repair and maintenance work on our elderly boat, I found myself working sixteen-hour days, seven days a week. It's good that I was young.

The quiet routine went on, week after week, interrupted occasionally by fire drills, battle stations torpedo, and battle stations missile exercises. But I was making progress. It was often repeated to me that duty aboard a nuclear submarine meant "many days of boredom and minutes of panic."

Mental relief came in the form of Sunday movies in the wardroom and crew's mess and family grams from Linda. Each sailor was allotted five family grams per patrol, and they were coveted. News from home and *I Love Yous* were the greatest morale boosters.

An SSBN on patrol almost never sends out radio traffic. It would alert the enemy to our presence and our location. Only in an emergency would we actively transmit. But we received daily very low frequency (VLF) transmissions on a trailing wire. Wives, girlfriends, children, and parents can write brief, officially monitored, and sometimes censored telegrams. Pornography, Dear John letters (i.e., *I filed for a divorce today*), or extremely disturbing news were not routed to the crewman.

Christmas at Sea

That year, *George Washington* Gold ended up with the Christmas patrol cycle. We would be at sea all of December. Somehow our cook found out it was my birthday. On December 22 he made an elegant baked Alaska and brought it into the wardroom. It was an act of kindness that nearly brought me to tears and certainly brought enjoyment to all *Washington's* officers. December 25th was my first Christmas Day at sea. Several sailors hung decorations, tidings of comfort and joy.

Most members of the crew and I quietly celebrated Christmas by opening presents our loved ones had put in our seabags. Linda had thoughtfully wrapped presents, including a beautiful photograph of her and our three daughters. It was strange and deeply sad to be hidden at sea on a day when much of the world celebrated with happy family gatherings.

Finally my intense focus and labors were rewarded. Near the end of patrol, just before we returned to the Rota Naval Base, I completed my walk throughs, signoffs, and board exams. I qualified as officer of the deck!

Standing my first OOD watch by myself, without being questioned or instructed, felt like I had just climbed Mount Everest (Earth's highest mountain).

As we were passing submerged through the Straits of Gibraltar, this time heading out, I happened to have the OOD watch. Inconveniently, we needed to go to periscope depth in the straits for a satellite navigation fix. Sonar was full of surface ship noises. As the scope broke the surface, I made a quick 360-degree scan. It looked like Times Square: multiple ship lights, lights from North Africa, even some lights from Gibraltar. The straits looked like the Ventura Freeway with dozens of ships going in every direction.

Captain Fields, standing next to me asked, "What do you see Dave?"

"Lots of lights, Captain. Would you like to take a look?"

Gratefully I turned the periscope over to the captain who stayed on the scope until we finished receiving the satellite signals and returned to a deeper and safer depth.

Second Off Crew

The Submarine Squadron notified *George Washington* Gold families of our return time and date. Linda and our girls met me at the Charleston airport. Their faces and embraces were heavenly. It was a comforting and loving homecoming after our longest separation.

Eager for news from my family, I asked how their weeks went. Linda said it felt like I had died. She had not heard from me for sixty-two days. That brought home to me one of the many sacrifices that submarine sailors make. There is no opportunity to call home. No letters written back and forth. It's as if you are dead.

Linda, one of the most thoughtful people I have ever known, had a decorated Christmas tree in our living room with colorfully wrapped presents underneath with my name on them. She made Christmas in January for me.

At home I experienced culture shock again. Desperately, I needed to make a mental transition from being in charge, making quick decisions, communicating with twelve words or less (with at least one profanity), and completing rapid information processing. What I needed to do was to intently listen to my young daughters' stories and immerse myself back into their lives, their school, and their neighborhood interests. It was a difficult mental transition.

Linda frequently, mostly gently, reminded me that I need not be in charge. Over time I learned that she had our family system on automatic, and all I really needed to do was take our daughters to the playground, give them baths, and read some bedtime stories. Oh, and take out the garbage, and make a few trips to the candy store.

After two weeks leave, I reported to the ship's office. Lifting a lighter load, I was able to concentrate on writing enlisted evaluations, help my sailors with their training and advancement, and study for my "last hurdle," achieving submarine officer qualification (my gold dolphins). Proudly, I added the Polaris Patrol Pin to my uniform.

What did I learn?

- ✓ Accomplishing difficult goals is hard work. But it can bring a sense of pride and self-esteem.
- ✓ The military lets you wear your accomplishments on your chest.
- ✓ An important role of a junior officer is to help your superiors succeed. "Make them look good."
- ✓ "Of all the branches of men in the forces, there is none which shows more devotion and faces grimmer perils than the submariners." —Sir Winston Churchill.

CHAPTER 22

A COLLISION AT SEA WILL RUIN YOUR DAY

Submarine Dolphins Insignia

After we relieved the Blue Crew to take over *Washington* for our second Gold Crew patrol, I focused intensely on my final signoffs and walkthroughs with the executive officer, commanding officer, and finally the squadron commander. It was finished! I earned my gold "dolphins." Qualifying in submarines as an officer was one of my life's most-difficult undertakings and one of my most-rewarding accomplishments, affirming that I had demonstrated a vast knowledge of all systems, operations, and casualty control aboard a submarine.

Having the confidence and foresight, or the audacity, to anticipate this, I had purchased my gold dolphins at the Navy Exchange before taking the crew flight to Rota. Opening my stateroom locker, I took them out and pinned them in place of my silver dolphins which I had been wearing for seven years, since my time on *Seawolf*.

No time to celebrate. We needed to get the Weapons Department ready for sea.

For the first time in my young, commissioned career, I had an assistant officer. He was a young ensign (most ensigns are young) whom I appointed as deck officer. That gave me more time to manage the torpedo and missile departments.

When my boss, the weapons officer (WEPS) had the duty, I took the opportunity to get off the boat with my new assistant. Planning to treat him to a dinner in Seville, we left the ship and approached the town. Suddenly we were met by a parade. When a huge, brightly lit float carrying three kings passed by it dawned on me that it was January 6th, the Christian Feast of the Epiphany. It was Three Kings Day in Spain. The city was packed, and the restaurants were full. We reluctantly returned to the base.

Being the newly minted officer of the deck and qualified in submarines ship's officer on *George Washington*, Captain Fields gave me the honor of getting her underway from Rota for our second patrol. The Bay of Cadiz was calm, and it was a pleasant, sunny day. I felt the warm air and sun on my face, inhaled, and savored the smell of the ocean breeze and had that "on the summit" feeling again. The US Navy had given me great responsibility and entrusted me with the safety of a 6,000-ton, multimillion-dollar, nuclear submarine, and 120 sailors.

The sight of a Soviet intelligence trawler dead ahead broke my sense of euphoria. The Russian ships were designated by the US Navy as auxiliary general intelligence (AGI) vessels. During much of the Cold War, Soviet AGIs gathered intelligence outside our ports and bases, and also harassed our ships and submarines. The Russian vessel was clearly out to harass us.

In my mind, I controlled my sense of outrage and calmly reviewed the rules of the road: *Just maintain your course and speed and pass safely by*. But the Russian had other plans.

The AGI began to get in my way. It passed back and forth over my intended course. To avoid a possible collision, I ordered, "All Stop," once and slowed down.

Soviet Intelligence Trawler (AGI) (National Archives #K-93182) courtesy Naval Historical Foundation. http://northernfury.us/warsaw/su/navy

I called down to the navigator to take a look through the periscope at the annoying ship and tell me what it was. He looked it up in *Janes Fighting Ships* and reported to me that it was a Soviet Primor'ye class intelligence trawler around 4,000 tons, 300 feet long, and 24 foot draft. She was powered by two diesel engines with a maximum speed of 12.5 knots.

The Soviet ship continued to play "chicken" with us. Several sailors on her deck were smiling, waving at me, and taking photographs. In my mind, I imagined a weathered face, seasoned Russian captain looking at me through his binoculars, laughing, and saying to his helmsman, "There's a kid driving that submarine."

What I was doing wasn't working, so I picked up the bridge microphone and said, "Captain to the bridge."

Captain Fields appeared within minutes and asked, "What's going on Dave?"

I explained the tactical situation and asked for his advice.

His advice was to call up flank speed, out run the Soviet trawler, and then sound the diving alarm from the bridge as we approached the 100-fathom curve. A fathom is six feet. We required 600 feet under the keel to safely dive the boat.

Nuclear submarines are designed and constructed to go at maximum speed when submerged. They don't go very fast on the surface. At flank speed, our bow wave was parting the ocean like Moses at the Red Sea. The AGI began to follow us, however it was falling behind. We were putting some distance between us.

I ordered the lookouts and all bridge equipment below, asked the navigator to inform me when we reached the 100-fathom curve, and notified the chief of the watch to prepare the ship for dive. As directed by the captain, I would sound the diving alarm from the bridge. That was an uncommon procedure, so I had to repeat the message twice. World War II fleet submarines routinely did crash dives from the bridge. Nuclear submarines did not. Submerging was normally a slow and deliberate evolution.

When we reached water deep enough to dive, the Soviet vessel was about 1,000 yards astern of us. I sounded the diving alarm, proceeded down the ladder, closed and secured the upper hatch, then the lower hatch, assumed the conn, and took the periscope.

The diving officer, who was directing the dive, reported, "green board" (telling me that all openings to the sea were closed.) I fixed my gaze on the fast-approaching, menacing, enemy ship.

Standing next to me, the captain informed me that the diving officer (per ship's procedure) had reduced the ordered bell to "Standard."

"Ahead Flank. 20 degree down bubble," I ordered.

OOPS! Too much down angle. Our propellor was coming out of the water. The prop was big, and visible, and I became further concerned about our ship's safety.

"Reduce angle to ten degrees down"

"Make your depth six zero feet"

"Down periscope."

As we began to level off at sixty feet, we could all hear the frightening propeller noise through our hull as the Soviet trawler passed right over the top of us. But we had managed to put enough water above us. There was no collision. No dreaded metal scaping sounds or damage to our sail or masts. Greatly relieved, I looked around at the many anxious faces in the Control Room and said, "Well done."

Next I requested a recommended course, speed, and depth from the navigator and directed the messenger of the watch to inform my relief that I was more than ready to be relieved. The confidence the captain placed in me that day was overwhelming. It is one of the most memorable experiences in my many years at sea.

The hours of weapons make ready, arming and certification procedures were now routine, familiar, and a bit easier. With my multiple submarine qualifications completed, I was looking forward to a less labor-intensive patrol. And no Wild Game Nights. But in mid-patrol we would receive unfortunate news. And it was devastating.

In addition to a Navy corpsman, ballistic missile submarines, at that time, carried a medical doctor. Our MD called a ship's officers' meeting to inform us that Captain Fields had cancer, was receiving care and pain medication, and was resting in his cabin. "Do not disturb him. The executive officer would be acting commanding officer."

We completed our patrol, again in the Mediterranean Sea and headed back to Rota. At Naval Station Rota, we evacuated our captain to the base hospital. His temporary replacement was a command qualified commander from Submarine Squadron 16.

The new CO took us back across the Atlantic Ocean to Charleston Harbor. At top speed, it took us ten days. We returned to the United States because *Washington* had been reassigned to the Pacific Fleet and would be based out of Pearl Harbor, Hawaii. But first we needed to pass another Operational Reactor Safeguard Examination.

We entered Charleston Harbor, while I was the officer of the deck. A tugboat met us and transferred the ORSE inspection party to our deck. We proceeded back to sea for the examination. For a second time the Gold Crew engineers failed their inspection. We were informed that we would remain on the boat for a six-week "training" period, then be reinspected.

Our homecomings were bittersweet. We would be with our families and loved ones again, but we would not turn over to the Blue Crew. We would keep the boat.

We got underway many times for training and drills. In and out for weeks with limited time ashore. As one of two weapons duty officers, I had in-port duty every other day.

Underway again, steaming down the Cooper River toward sea, I was on deck, supervising the deck crew. My family was on the shore, watching us leave. I could hear my daughters mournfully yelling, "Daddy. . . Daddy." My sadness was heavy.

By the time the Nukes passed their inspection, we "owned" the boat for nearly five months. Not long compared with many other US Navy ship's operations, but it was unusually long and arduous for an SSBN crew. And it was hard on the Gold Crew. Never had I seen such low morale and open discontent in a submarine crew before.

What did I learn?

- ✓ When you think you are over your head. You probably are.
- ✓ When you think you need help. You do.
- ✓ Seek help from people with more experience than you.
- ✓ Ask for advice and share responsibility for difficult decisions.

CHAPTER 23
PASSED OVER

Death of a Captain

We buried Captain Fields, in summer that year. His funeral was held at an old, small Episcopal Church in Charleston. He had not reached his fortieth birthday. He left behind a widow, Janet, and three young children. The Navy lost a fine officer, and I lost a friend and a mentor. He helped me progress in my seamanship and leadership. After my father, he was the second man's death I grieved and damned the curse of cancer.

Fitness Reports

For most of my Navy career, I had received favorable enlisted evaluations that accurately reflected my good job performance. Recommendations for promotion and assignment to positions of greater responsibility were common. As a result, I was rapidly promoted.

Consequently, I became very adept as an officer at evaluating and counseling sailors in my divisions. I took a sincere and active interest in the training, welfare, and advancement of my sailors. "Take care of your men and they will take care of you," was my belief.

The officer equivalent of the enlisted evaluation is the Officer Fitness Report (FitRep). I had paid little attention to the few perfunctory reports I had received from my previous Navy schools and brief assignments. But I was taken aback by the Fitness Report signed by my commanding officer while in the Charleston Naval Shipyard in 1971. It was accurate, but turned out to be damning.

It recorded that I was having trouble keeping up with all of my duties: personnel management, qualification as a nuclear weapons duty officer, launcher officer, first lieutenant, diving officer, officer of the deck, submarine officer qualifications, and shipyard repairs. But I was incensed that I was ranked seventh out of seven officers on *Washington*.

I felt that I was busting my butt while the Nukes sat in the office all day studying their ship's nuclear propulsion manuals (RPMs). And the Gold Crew Nukes could not even pass their inspections, while I passed all of mine. In my meeting with the commanding officer, he told me that I was not meeting performance expectations. I should have left it at that. But I didn't. Instead I chose to write a letter to the Navy Personnel Command challenging my ranking.

The executive officer warned me that it would be a mistake. He was right. I screwed up. It was an error in professional judgment that ultimately hurt my naval career.

So I was not a surprised when I received the news that I had "failed selection" for Navy lieutenant in 1972. But I was shocked and saddened. When I was informed, I was also counseled that I would now move to an "above the selection zone" status that would make my chances of advancement even slimmer. This is not how I envisioned my officer career. It would be cut short, and I would not "assume positions of greater authority" in the active Submarine Service.

Since I had just over a year left in my now-abbreviated commissioned officer service obligation, I was given the option to stay aboard *Washington* and go to Pearl Harbor, Hawaii, or be transferred to another Charleston-based submarine.

Moving my family and finding civilian employment after separation was going to be hard. It would have been doubly hard to do it from an island in the Pacific Ocean, so I chose to be transferred.

Prior to my departure from *Washington,* Linda and I were presented with a silver serving bowl. The thoughtful gift was inscribed with words of gratitude for my years of service aboard the USS *George Washington* SSBN 598.

That summer, Linda and I had purchased a small sailboat. One warm, pleasant day, we launched our boat into Charleston Harbor. We were enjoying a day sail when the rudder suddenly came loose and floated away. We lost control of the boat and capsized. We found ourselves in a life-threatening situation. Charleston Harbor was known to host sharks and is not a safe place for swimmers. We were both very concerned for our safety. But at least my life was not passing before my eyes!

Linda and I are both strong swimmers, having grown up surrounded by the Great Lakes. Fortunately, we were rescued by people in a motorboat who towed our swamped boat to shore. We sold our small, rudderless sailboat the next week.

What did I learn?

- ✓ "The most important factor in sailing with safety is the condition of your boat. This demands eternal vigilance, constant inspection and proper maintenance." —Hervey Garett Smith, *The Small-Boat Sailor's Bible*
- ✓ Take time to manage upward.
- ✓ Remember New York Mayor Koch? Mayor Koch walked the streets of New York City, asking New Yorkers how he was doing. But you don't have to be an elected official to ask for feedback. In fact, it's good for your career to ask your manager/supervisor/boss, "How am I doing?" When you do, you are likely to get some helpful information and advice. Use it to make necessary adjustments in your job performance.
- ✓ Especially in our free-enterprise, employment-at-will culture, your "superiors" have immense power over your pay, promotion, and livelihood. Your job security is often in the hands and mind of your immediate supervisor who writes your performance evaluations.
- ✓ Frequently tell your boss what you are doing, and ask, "So what do you think?"
- ✓ All actions have consequences. Rash, emotional actions can have unfortunate consequences.

CHAPTER 24

LEWIS AND CLARK

USS *Lewis and Clark*, SSBN 644
courtesy NHHC catalog #L45-164.05.02;
USS *Lewis and Clark* medallion,
courtesy NHHC, catalog #NH 64724-KN

Poseidon

It would become my lowest mental point in the Navy since my days as a mess cook. I needed to cope with professional rejection, loss of my dream career, and more separation from my family. Two weeks after departing *Washington* after many months at sea, I drove north to Dam Neck, Virginia, for six weeks of Poseidon Missile Officer School. I left my family behind in tears.

My new submarine was Charleston-based USS *Lewis and Clark*. She was named after American explorers, Second Lieutenant William Clark and Merriweather Lewis. They were commissioned by President Thomas Jefferson in 1804 to explore lands west of the Mississippi. The exploration lasted more than two years traveling over approximately 8,000 miles of treacherous waters and unforgiving terrain. Their efforts provided valuable geographic, ecological, and social information about formerly uncharted areas of North America.

At the time, *Lewis and Clark* was one of our newest missile submarines (commissioned on December 22, 1965). And she was big: 425 feet long (about 1.5 football fields), she displaced 8,000 tons and carried sixteen Poseidon missiles. The ballistic missile was named after the Greek God, Poseidon, God of the Sea and protector of seafarers. That's why I needed to go to Poseidon Missile School in Dam Neck, Virginia.

Like the submarines that carried it, the Poseidon missile was huge: 64,000 pounds, 34 feet tall, with a two-stage, solid-propellent rocket engine, with a range of 3,000 nautical miles. Poseidon was the leading edge of our nuclear deterrence, capable of carrying up to 14 RVs (reentry vehicles, i.e. nuclear warheads.)

Assigned as assistant weapons officer (AWEPS) on the Gold Crew, *Lewis and Clark* was my fourth boat. After graduating from missile school, I reported aboard to the off-crew offices.

My new captain, Commander John Gordon McGarry, was predicably a naval academy graduate and a nuclear-trained officer. He was unexpectedly affable, personable, and warm.

To get up to speed on patrol planning, I set about to meet my weapons crew members. We would depart for Rota in less than a month. My new commanding officer gave me a short list of qualification requirements I needed to complete in order to stand watch on his boat. In addition, he appointed me communications officer and asked me, as torpedo officer, to prepare for torpedo firing exercises to be conducted once we took over the boat.

The contrast could not have been more stark, or more welcome. After serving on a boat beset with mechanical and morale problems, tenseness, and stress, *Lewis and Clark* was high-spirited and low stress. It was a happy crew.

The evening before our crew flight to Spain, Linda reminded me that in the last year, I had been away from home for nearly eight months. Unusual for two-crew, missile submarine duty.

Unable to find any consoling words, I could only express my admiration for her abilities to "keep it together." Eventually the *Lewis and Clark* "wardroom wives" surrounded her with support and friendship.

My new boss, Jerry, the weapons officer, was also a NESEP graduate. He immediately demonstrated confidence in me and empowered me to manage my three divisions and forty crew members. My extensive experience on *Washington* more than prepared me for my new assignment. I felt like a combat veteran.

After the change of command in Rota, I was tasked with preparing two exercise torpedoes for the at-sea firing exercise. Captain McGarry, the commanding officer, would act as the approach officer responsible for developing the firing solution. *Lewis and Clark* would fire at a surface practice target. We fired two exercise torpedoes at a target tug. My next job was to reconstruct the accuracy of the shots.

Fortunately, the commanding officer was on target, and I didn't have to report to him that he missed. *Lewis and Clark* only had four torpedo tubes forward, so the make-ready procedures were easier, but the missile warhead closeout operations were just as time-consuming.

My requalification went well. I was soon standing underway watches as officer of the deck. Although we had several routine battle stations and simulated casualty drills, this sixty-two-day patrol was, hands down, the most relaxed period in my years of sea duty.

Accustomed to serving on old boats, I kept waiting for mechanical and repair problems to surface, but few did. *So this is what a new ship is like.*

The Communications Department, my added responsibility, was easy additional duty. The radiomen required little supervision. I got to read and review all ship's radio traffic and many family grams. But to reduce risk, I insisted on a complete inventory of all secret and top-secret publications. Trust but verify.

Not having a substantial work burden and the anxiety of qualification requirements actually took a long mental transition. Accustomed to a fast pace and a heavy workload, I eventually settled into a more-relaxed routine. Only twelve-hour workdays rather than sixteen hours.

For the first time on a deterrent patrol, I took an interest in the Navy Ballistic Missile Submarine tradition of "halfway night." Halfway night, when the patrol is half over, the officers and crew plan a party with entertainment including music and talent shows. Many submarine crew members played musical instruments and had musical talents. Of course, since alcohol is never allowed on US Navy ships, it's an alcohol-free party.

I planned and organized a college bowl-type competition between the Engineering Department and the Weapons Department. Quiz questions were taken from the standard submarine qualification books. The event was a success. It was so well received that it was mentioned in my next fitness report.

Having spare time for the first time, I took up the study of naval history. Since we were on patrol in the Mediterranean Sea, I again viewed much of the Med through the periscope.

My frequent, silent lament was, *It would have been great if the Lewis and Clark had been my first boat instead of my last.*

Christmas at Sea Again

It happened a second time. Just like *Washington, Lewis and Clark* Gold caught the Christmas patrol cycle. Once again, we all quietly opened our presents from home that had been stealthily brought aboard in a special trunk.

Saturday Night Poker

As a ships' officer with money, I was invited to join the captain for Saturday Night Poker in the wardroom. Gambling on Federal property is legal, and Navy ships are federal property. The dollars won (or lost) were recorded by the captain in a green book he labeled the Laundry Bill.

Early in the patrol, I was winning. However, my fellow officers discovered that I bluffed often. After a bet or a raise, at least one person at the table shouted, "Mr. Francis is bluffing again." By end of patrol, I had a modest laundry bill to pay.

What did I learn?

✓ You can get away with bluffing others for only a short time.

✓ Being a ship's officer is hard work, but it can also be enjoyable and rewarding.

CHAPTER 25
INTO THE STORM

Beaufort number 10
Description whole gale
Wind speed 55–63 mph
Wave height 44–59 feet

Sea conditions

"Very high waves with long overhanging crests; resulting foam in great patches is blown in dense white streaks along the direction of the wind; on the whole the surface of the sea takes on a white appearance; rolling of the sea becomes heavy and shocklike. —Navy manual, *Knights Modern Seamanship*, Fourteenth Edition, D. Van Norstrand Company, Inc.

Watercolor by Henry Reuterdahl; C. 1919, courtesy NHHC #60-380-V

"The darkness palpitated down upon all this, and then the real thing came at last. It was something formidable and swift . . . It seemed to explode all around the ship with an overpowering concussion and a rush of great waters." —Joseph Conrad, *Typhoon*

"It's nasty weather like this here that turns heads gray before their time."
—Jack London, *The Sea Wolf*

It was a routine refit in Rota. This was my second run on the *Lewis and Clark,* and my last SSBN patrol before I left active duty. In honor of my last underway, Captain McGarry gave me the bridge watch. Unfortunately, there was a gale raging in the North Atlantic Ocean. But *Lewis and Clark* could not delay our departure. We were obligated to get underway and to be on station, on time, to cover our assigned target package .

If you've seen the movie, *The Perfect Storm,* you have a vision of the sea conditions. Standing on the bridge, I drove the ship into the storm, into a rush of violent waters, an angry, agitated ocean, mountainous waves, and a menacing, unnatural green sky. It was an awesome, frightening sight. After sending the lookouts below with orders to secure the bridge hatches, I fastened the chain attached to my safety harness to the side of the sail and braced myself for a wild ride.

Riding on the savage sea, the ship slowly climbed up the front of each giant wave, careened over the crest, then slid rapidly down the backside into the trough where the ship shuddered violently. Each time we hit a trough, I was assaulted with gallons of cold, green ocean water. The bridge would flood up to my waist then drain. Then we began to ascend the next wave. It was a scary roller coaster ride, and I was seriously motion sick.

In one of the troughs, the shaking was especially strong. I again ducked down in anticipation of the next deluge of water. When I stood up, I saw that the plexiglass windshield, which offered no protection in a violent storm, had shattered. If I had not ducked, shards of glass and pieces of steel like shrapnel would have hit me in the face, neck, and chest. I was lucky, wet, cold, and sick.

On my order, the messenger of the watch came up to the bridge with a set of wrenches. Together we unbolted and jettisoned the remains of the windshield, which had continued to threaten me by flapping in the wind in front of my face. Word from the control room that we had reached the 100-fathom curve (water deep enough to submerge) sounded like a love song.

Following submerge the ship procedure, I went below, secured the upper and lower hatches, and took my officer of the deck position on the periscope stand, dripping wet and shivering. "Green board" was reported to me. The captain greeted me with a broad smile and gave me permission to dive the ship.

My submerged officer of the deck watch relief arrived and so did the navigator who proceeded to scold me for not saving the metal frame of the windshield.

"Navigator, your *#%#XQ&#* windshield almost killed me," I protested.

During my last months at sea as an officer aboard a nuclear missile submarine, I was a trusted senior-watch stander, and division officer to four divisions comprising forty highly trained submarine sailors, one-third of the crew, and responsible for the safety and security of thermonuclear missiles and torpedoes. It was the outcome of many years of resilience, study, persistence, discipline, toil, tears, heartache, and help from many along the way. Since my beginning as a seaman recruit, I had climbed many steps on the Navy's ladder and felt pride in my achievements coupled with sadness that I would have to leave active duty that summer.

What did I learn?

- ✓ We will encounter roughs seas. Face them with foresight and courage.
- ✓ A smooth sea never made skilled mariners. (English Proverb)

Some went down to the sea in ships
and plied their trade in deep waters;
They beheld the works of the Lord
and his wonders in the deep.
Then he spoke, and a stormy wind arose,
which tossed high the waves of the sea.
They mounted up to the heavens and fell back to the depths;
their hearts melted because of their peril.
They reeled and staggered like drunkards
and were at their wits' end.
Then they cried to the Lord in their trouble,
and he delivered them from their distress.
He stilled the storm to a whisper
and quieted the waves of the sea.
Then were they glad because of the calm,
and he brought them to the harbor they were bound for.
Psalm 107. *The Book of Common Prayer*

CHAPTER 26
MINNESOTA ON THE PRAIRIE

The Francis family had adjusted to the routine of a seagoing sailor's life and a "well oiled" Navy family. But now, as Linda said, "It's fruit-basket upset." Secure on the government payroll for over ten years, 1974 was a year of economic recession, and the first year I started to pay attention to the health of the US economy. Not the best of times to find a "civilian job." At the time, we were in the throes of the deepest post-World War II recession. The combined high unemployment with high inflation was labeled "stagflation." The Federal Reserve Bank of Minneapolis described the economy as unsettled, "with a possibility of rough seas," due to interconnected issues of inflation, recession, the energy crisis, and questions of financial stability (domestic and international; long-term and short-term).[10] (I was accustomed to rough seas.)

The transition office at the Charleston Naval Base offered little career knowledge or help. After months of my individual job search effort, the stack of corporate rejection letters was getting higher. I was optimistic that my old employer, Dow Chemical in Midland, Michigan, would be interested. But they were not. Apparently, my BS in chemistry was not turning heads.

---

10 "The Limping Giant: The American Economy: 1974-75" by Bruce K. McKlaury. The Federal Reserve Bank of Minneapolis, January 1, 1975. https://www.minneapolisfed.org/article/1975/the-limping-giant-the-american-economy-197475

In the midst of my self-doubt and growing anxiety that I would fail at supporting my family, I received a call from the 3M Company in St. Paul, Minnesota. I had never been to Minnesota, but knew it was somewhere west of Michigan, and it was a cold place. The Human Resources rep invited me for an interview. They would even pay for my travel and lodging.

I was glad to learn that Minnesota had a fairly diverse economy with a standard of living and level of education among the highest in the US. Growing up with my family in the hardware business, I was familiar with 3M (Minnesota Mining and Manufacturing) and was aware that it was a fine company. Founded in 1902 as a small mining operation in Northern Minnesota, 3M extracted corundum, a mineral for making abrasives. Though the initial venture failed, the company's ongoing innovations in science and technology eventually made it a multibillion-dollar, international manufacturer of thousands of products used in homes, business, schools, hospitals, and industry.

In early April, it was spring and 85 degrees Fahrenheit in Charleston as I flew out to Minnesota. When we landed, the local temperature was in the 30s, the sky was gray, and the ground was brown. *Well, welcome back to the northern climate.*

As directed, I took a cab to a motel near the 3M Center. The driver took the Mississippi River Road where I noticed stone bluffs, forest, and a wide river (the Mississippi) which reminded me of New England. From my motel room, I called 3M Human Resources to report my arrival. She said I would be picked up in the morning for my interview.

April 4, 1974

The next morning I was greeted by Mike Miller and Al Anderson, managers in the Safety Systems Division. Safety Systems manufactured reflective products for highway signs, road markings, and license plates. You see them everywhere. The position they wanted to fill was a process engineer. Process engineers at 3M work to improve manufacturing processes and reduce production costs.

Without asking, I learned early in the conversation why in the world they might be interested in me. Al Anderson was a former submarine officer and a lieutenant commander in the Naval Reserve. He was attracted to my resume because I was a (fellow) submarine officer. He explained that even though 3M prefers chemical engineers, he thought my technical degree coupled with my submarine engineering training gave me the necessary, practical, industrial engineering skills. I was relieved and encouraged to hear that.

Later that day, Al's wife Carol joined us. Carol and I commiserated about the long separations and strains on the family of a seagoing submarine officer. I felt comfortable with Al and Carol. *This might just work out.*

It began to rain that morning, turned to freezing rain, then to snow. Al apologized to me often throughout the day for the "unusual" weather. The Twin Cities airport closed due to the ice, so I was dropped off at a St. Paul hotel for the night. I called Linda to share the day with her.

She asked, "What's it like up there?"

"It reminds me of New England. You will love it."

Back home in South Carolina, Al called me with a 3M job offer. Linda and I were elated that we had navigated a difficult transition. Afterward, I received an interview request from General Electric. Good for my morale, but too late to consider.

Next up in May was a one-week, all-expense paid trip to Minnesota to find a house. I was beginning to like the corporate world.

Farewell

Gordon McGarry and his wife Joan hosted a farewell party for me and Linda and two other officers who were departing *Lewis and Clark*. We assembled for dinner in a hunting lodge outside of Charleston. It was a little unnerving to sit in a room surrounded by dead animal heads, but the food was delicious, and our dinner companions were warm. Linda and I were presented with gifts. Appropriately, some of the gifts were 3M products like Scotch Tape.

Two farewell gifts were especially thoughtful. The first was a copy of *In the Footsteps of Lewis and Clark*, a chronical of the Lewis and Clark Expedition, written by Gerald Snyder and published by the National Geographic Society. It was signed and filled with well wishes. The second gift was a silver ice bucket. Similar to the silver bowl we received from the *George Washington*, it was inscribed with words of gratitude for my service aboard the USS *Lewis and Clark* SSBN 644.

We asked my mother to watch our girls while we went house hunting in Minnesota in May. Linda had arranged for a realtor to take us around neighborhoods east of St. Paul near the 3M Center.

Linda and I encountered sticker shock as we came face to face with the higher housing costs in Minnesota compared to South Carolina.

We found one house we thought we could afford, and our offer was accepted. We had a new Minnesota home, but we needed to sell, not one, but two houses, in Charleston.

Linda did her magic again. Our renter in our Pepper Hill (North Charleston) house agreed to buy it. Next Linda put our west-of-the-Ashley house "For Sale by Owner." She was confident she could sell it. And within a few weeks she did!

Lost and Damaged

Our move to Minnesota was our eighth move in ten years. 3M offered to pay our moving expenses, but I declined. We would use our last US Navy benefit, a final move, and save my new employer some money. It was a mistake. Our July relocation started and ended badly.

The moving van in Charleston showed up with a partial load. There was not enough room for all our household goods. Two thousand pounds of our belongings went to a temporary storage facility, to come later.

We drove to Minnesota in our two cars. Linda and the girls traveled in our station wagon with me trailing behind in our second car, my VW Beetle. Yes I bought another VW. It was affordable. We closed on our house, then called North American Van Lines to deliver our furniture. They could not find the truck or the driver. They were both missing.

I reported for work each day in the same sincere, brown suit and called the moving company every day, demanding that they find our belongings and return them to us. After two weeks of daily calls, I finally got an answer. The moving van driver had developed a case of chicken pox, parked his truck at his home in Spartanburg, South Carolina, and went off the grid. It was a cause for celebration the day the van arrived with our furniture.

The temperature was over 100 degrees Fahrenheit in July when we moved into our new Minnesota house, which lacked air conditioning. Linda and I had previously moved seven times in ten years. This eighth move was, by far, our most troublesome. It took us several weeks to put our lives back together.

The following January, we experienced our first Minnesota blizzard. Deep snow (27") piled up outside our windows, the temperature fell to 10 degrees below zero, Fahrenheit, 60 mph wind gusts created a minus 80-degree wind chill. Though Linda and I both grew up in Michigan, we had never seen weather extremes like that. It was a 110-degree swing from July to January. During that blizzard, I turned to Linda and said, "What are we doing here?"

From January 9 to 12, 1975, that blizzard, later referred to as the "Super Bowl Storm of '75" by the *Farmers' Almanac*, affected most of the Central and Southeast US. The storm produced forty-five tornadoes in the Southeast, dropped more than two feet of snow in the Midwest. Fifty-eight people and more than 100,000 animals were killed. This storm remains "one of the worst blizzards to ever strike parts of the Midwest, as well as one of the largest January tornado outbreaks on record in the United States."[11]

My new neighbor told me that Minnesota has a climate similar to Siberia. So that explains the weather extremes. In addition, we discovered that our new adopted state had substantial income and property taxes. Of course, they support a high quality of life and education. We had been exempted from income tax and paid low property taxes in South Carolina, so our new civilian family budget was very tight.

Our three young daughters adapted quickly to their new environment. In the fall, Robin and Jocelyn went to nearby Castle Elementary. Linda enrolled Melissa in the Muffin Man Preschool. And we began attending a small Episcopal Church, a welcoming faith community only three blocks from our house.

When we arrived in Minnesota our girls spoke with a beautiful, Charleston, southern drawl and addressed adults as either Sir or Ma'am. Those lovely southern traits disappeared in about six months.

---

[11] The Super Bowl Storm of '75" by Jaime McLoud. *The Farmer's Almanac*. Dec. 5, 2020 https://www.farmersalmanac.com/the-super-bowl-storm-of-75

What did I learn?

- ✓ Based on my personal experience and the stories of other veterans, the transition from active military to civilian life and employment is one of the most difficult.
- ✓ I've read that military transition services have improved. In addition, to pull it off successfully requires lots of homework, study, and forethought.
- ✓ Transition discussions with family, friends, and neighbors are helpful.
- ✓ Leaving active duty can be a good time to go back to school for skills training or higher education.
- ✓ Leaving active duty? Do your homework, Read books on job search strategies and resume writing. Research the private sector, government or nonprofit careers.
- ✓ Seek expert advice.
- ✓ No one moves to Minnesota or stays here for the climate. We stay here despite the climate and the taxes, for the quality of life that the state offers.
- ✓ There really is a Minnesota Nice. While in high school, our youngest daughter Melissa wrote an award-winning short story about Minnesota entitled, "Cold nose. Warm heart."

CHAPTER 27

LIEUTENANT, UNITED STATES NAVAL RESERVE

Image courtesy of military-ranks.org.

My manager at 3M, Al Anderson, a Naval Reserve lieutenant commander, encouraged me to join the Naval Reserve. It made sound financial sense. With my thirteen years of service, I could retire with a small pension in seven years. So in August, I joined the Naval Reserve. This was full circle for me since I began as a naval reservist. Shortly after I signed my Naval Reserve papers, I received a letter from the Department of the Navy: my promotion to lieutenant, United States Naval Reserve. Lieutenant is the third commissioned officer rank in the United States Navy, comparable to the rank of captain in other branches of the Armed Services.

Once again I learned to appreciate the important role of a mentor (wise advisor, teacher, and coach) in my life and career. Al, a superb manager and leader, was selected as commanding officer of a Naval Reserve cruiser unit. He appointed me his officer in charge of unit training. This was my first paid reserve officer assignment. Just what I needed to keep the family afloat.

The Naval Reserve had changed significantly since my enlistment in 1961. Back then it was like a secretarial pool. We were a large collection of sailors not knowing where we would end up if we were mobilized in a reserve call up. By 1975 each reserve unit had a sponsor, an active-duty ship or station participating in our training, and the place to go if we were mobilized. The reason military reserves exist is to augment the regular forces—to add strength by increasing the number of trained service members to respond to a national emergency or global conflict.

This "augmentation" concept would become more widely known when our national guard and army reservists were activated, often for multiple tours of duty, in Iraq and Afghanistan.

As training officer of US Navy Cruiser Reserve Unit (CG 216), I was given the opportunity to apply my experience conducting training for others. Our sailors were not submarine sailors. They were mainly surface sailors, as we called them, but they were eager to learn, advance and perform. My commanding officer empowered me to organize the members into groups with common training requirements. In addition, I revived the old concept of allowing time and encouraging self-study in their rates (specialties). There were multiple benefits to the sailor and to the Navy by advancing in rate.

It provided greater value to the service, increased morale, pay, promotion, and leadership opportunities.

Self-study hours also gave the training staff additional time to prepare lectures. In my first reserve officer assignment, I felt confident in my abilities. I drew upon my skills and experience from my active-duty years as a sailor and as a ship's division officer. It was like riding a bike and I was highly motivated to do well.

Using my past experience selecting competent leaders, I cobbled together a capable training staff and appointed instructors. It was coming together well. Soon we had excellent weekend training schedules, documents, and evaluations.

The Reserve Unit Readiness Rating is the coin of the realm. It is the ultimate and complex metric that shows how ready a reserve unit is to be mobilized and integrated to support an active US Navy ship or station. We trained to achieve the top rating, an R-1 readiness score.

Next requirement was to plan our first Active Duty for Training (ACDUTRA). This was something I'd never done before. *Okay. Read the manual and talk to people who have.*

Going to San Diego

For over ten years, I was an East Coast sailor. My orders, assignments, and ships had me and sometimes my family moving up and down the Atlantic Coast from Maine, New Hampshire, Rhode Island, Connecticut, Maryland, Virginia, South Carolina, Florida, and Rota, Spain. Never before had I been to a West Coast naval base. For the first time, I would see the Pacific Ocean. We were going to San Diego, California!

In September 1975 my reserve unit and I flew out to San Diego for two weeks reserve active duty aboard the USS *Long Beach*. *Long Beach,* a guided missile cruiser commissioned in 1961 was the US Navy's first nuclear-powered surface ship. This was the first time I had stepped aboard a surface vessel since my destroyer cruise in 1962. The big, impressive ship was 15,500 tons, 721 feet long, with two nuclear reactors and a crew of 1,160.

Long Beach was designed as an "all-missile" ship from the very beginning, but was fitted with two five-inch 38 caliber gun mounts amidships at President John F. Kennedy's order.

USS *Long Beach,* oil painting by A. La Paca; C. 1961, courtesy NHHC, accession #95-001-G.

She was a "storied" ship that in 1964 had steamed with the aircraft carrier *Enterprise* and the guided missile cruiser

Bainbridge as an all-nuclear-powered battle formation on an around-the-world cruise. *Long Beach* had also been extensively deployed in the Western Pacific during the Vietnam War. My training plan, approved by my CO and *Long Beach,* called for a ship-wide orientation for all officers and enlisted members from my reserve unit. In addition, I placed some reservists in local schools and in ship's divisions to get hands-on training and provide help to the ship's crew with their maintenance and upkeep work.

As an "old" weapons officer, I relished my introduction to the Terrier and Talos Missile Systems, Electronic Warfare, and Fire Control Systems. We reserve officers became familiar with the engineering steam plant (not the nuclear reactors), damage control organization and procedures, and the ship's standard operating procedures. We were all immersed in technology and always busy. In addition, I stood in port watch as assistant command duty officer.

We made it to San Diego Harbor one evening to dine at a seafood restaurant that required that we wear neck ties. We were tieless, so they offered us a wide variety to choose from. That evening, while enjoying the warm weather and great seafood, I called Linda back home in Minnesota and said, "Sell everything and meet me in San Diego."

She said, "No. You should come home after you're done there."

But I was not done with my West Coast Navy adventure. At the end of my workday, I asked for directions to the base officer's club. *Long Beach* was tied up at Coronado Island. The shortest route from the ship to the O Club was to walk across the Naval Air Station.

So I set off walking across the airfield. Off in the distance, I saw the lights of a fighter jet coming in for a landing. Running quickly, I made it off the concrete just as I heard the squeal of the jet's wheels landing behind me. And luckily, I was not apprehended by naval air station authorities. On my return trip to the ship, I chose the long way back. Not pressing my luck, I circled the airfield instead.

What did I learn?

- ✓ In April 1975, with the fall of Saigon, the US lost the war in Vietnam and evacuated in helicopters carrying 1,000 US civilians and 7,000 Vietnamese refugees. I learned the feeling of deep sadness and trauma at the losses—both in war and in the immense loss of life: more than 58,000 Americans, 1,350,000 North and South Vietnamese soldiers, and two million civilians on both sides.[12]
- ✓ Until the war in Afghanistan, the Vietnam War was our longest and costliest. Our Vietnam veterans did not come home to open arms, parades, or public applause, but to scorn, insults, and national neglect. It was sad. We in uniform were warned not to wear our uniforms in public.

[12] This Day in History, Vietnam War Timeline. Updated Feb 26, 2020. https://www.history.com/topics/vietnam-war/vietnam-war-timeline

CHAPTER 28

HAWAII

Insignia of Submarine Base, Pearl Harbor, courtesy NHHC, catalog #NH 69768-KN.

You can't make this stuff up. In 1977 the US Naval Submarine Base, Pearl Harbor, Hawaii, established a new reserve unit at the Naval Reserve Center in the Twin Cities of Minnesota. Recently promoted Commander Al Anderson was selected as its first commanding officer. LT. Francis was appointed as its first training officer. Our gaining command, where we would do our annual training, was in Hawaii! My travels to West Coast naval bases would soon expand to the Hawaiian Islands.

The Pearl Harbor Submarine Base is legendary in the history of the United States Submarine Service. Workers at SubBase Pearl

Harbor prepared and repaired our fleet submarines for war patrols in the Pacific Ocean during World War II. Their skills and labor helped win the naval war in the Pacific. On December 7, 1941, just six hours after the bombing of Pearl Harbor, the order went out to the American Pacific and Asiatic fleets to execute unrestricted submarine warfare against all Japanese shipping. Though it did not begin well due to a multitude of problems, American submarines eventually sank 55 percent of all Japanese ships lost during the war. The total included over 200 Japanese warships, among them two of the aircraft carriers that participated in the Pearl Harbor attack. The Silent Service suffered 40 percent of all naval casualties in the Pacific: 52 US Submarines, 374 officers and 3,131 enlisted men are on "eternal patrol."[13]

In 1977, Submarine Base Pearl Harbor cared for nuclear submarines fighting the Cold War. Our Midwestern reserve unit of seventy sailors joined this tradition. Commander Anderson attended a conference in San Diego, California, for reserve unit commanding officers, hosted by the commanding officer of the Submarine Base, Pearl Harbor. When Al returned, he announced that we would do our first two-week active duty in March. What a great time to go to Hawaii! March in Minnesota is a winter month when we begin to complain about the relentless cold weather that just won't go away. We asked Al how he got such a favorable date.

He said, "I happened to be seated to the left of the Submarine Base commanding officer and he handed the blank schedule to me."

---

[13] "America's Undersea War on Shipping." James M. Scott, US Naval Institute. December, 2014.

Ed Kimball, an enlisted submarine-qualified lieutenant, senior to me, was selected as executive officer. The three of us, Al, Ed, and I, met for a planning session. Ed was proposing a standard shipboard organization for our new unit: Engineering, Operations, Weapons Departments, etc.

Since we were not a ship replacement unit, but a repair facility unit, I suggested that our training should match the submarine base's organization. My shop-focused plan was adopted.

With that settled, I set about reviewing the service records of the chiefs and senior enlisted assigned to us. Since there were no other officers besides the three of us, we would rely heavily on our senior enlisted for team leadership. Lastly, we all three submariners resonated to the concept of appointing a senior enlisted advisor (SEA), a reserve unit equivalent to the chief of the boat.

Since this was long before personal computers and laptops, we drew out the organizational scheme with personnel assignments, in a form that could be displayed using a 3M overhead projector. It was a flat organization with heavy reliance on senior enlisted members leading specialty groups (teams) tasked with training and performance evaluations to support an Intermediate Level Maintenance and Repair Activity (IMA). It was not innovative, but at the time it was unusual.

Since both Al and Ed were excellent leaders and personnel managers, I could concentrate on the training program. Only two months after we launched the unit, we received a command inspection. We nailed it! Appearing in my next officer fitness report, again with a complete lack of modesty: "Very high marks were achieved in all inspection areas . . . The training plans and classroom training were noted to be outstanding. The inspecting officer noted that this unit was one of the best-organized units in the Region Sixteen Readiness Command." That report was welcome affirmation that I had gained some knowledge about organizational development and leadership.

Hawaii

Next we set about planning for our first Two-Week Active Duty for Training in March 1978 in Hawaii. Each of our spouses, Carol, Carolyn, and Linda, decided they would accompany us. We booked rooms at the Hale Koa (House of the Warrior) on Waikiki.

Several inches of snow were on the ground in Minnesota as we boarded the flight to Hawaii. Commander Anderson prescribed service dress blues for our travel uniform.

On the airplane, as I walked down the aisle for a break, a passenger stopped me. Mistaking me for a pilot, he asked me how fast we were flying.

"We're doing about 500 knots," I responded.

He was satisfied with my answer.

The next morning, we reported to the Administrative Office at the Submarine Base at Pearl Harbor. We were wearing our "salt and pepper" (white shirt and blue trousers) uniforms authorized in the Naval Reserve at the time. We were greeted by "Hairy Finger," the former executive officer of the USS *Bang* who was now executive officer of the submarine base.

It was *deja vu*. Just as he had ordered me to button my shirt pocket in 1965, he told us in 1978, that we were out of uniform. "Salt and Pepper" was not an authorized uniform on *his* base.

We all had serious cases of jet lag. This was my first trip to the Hawaiian Islands and it reminded me of my first crew flight to Rota, Spain, and my thirty-three hours without sleep. But I was able to function adequately, I believe.

Still in the process of building our training library back at the reserve center, I created a scavenger hunt for our enlisted members. Find and collect as many technical manuals as you can legally retain. They would become training curricula for our classroom use. We amassed a significant collection.

My typical training day had me alternating between managing a Pearl Harbor Repair Department Engineering Project and checking on the repair activities of our sailors. At the end of our workday, we three officers returned to the Hale Koa. We were tired, but our spouses were anxious to go out. They had time to do the restaurant research, so the six of us would walk to the restaurant selected by our wives for the night. We three officers began to refer to our evening routine as "forced marches."

You cannot be on the island of Oahu without ubiquitous reminders of "December 7, 1941, a date that will live in infamy," as President Roosevelt said when he asked the US Congress for a declaration of war against Japan. Reminders of the tragic loss of sailors and civilians, and the destruction of ships, aircraft, and buildings are ever present in memorials, ships' plaques, gravesites, and buildings that survived where people felt fear and anger and where they worked for years to fight back and achieve victory.

At the old Submarine Base Officer's Club, I was filled with awe and gratitude for the submarine captains and officers who walked those floors and who won the war in the Pacific.

Our small group, three officers and three spouses, became our new wardroom. We grew in affection, mutual respect, and thoroughly enjoyed being together. Future assignments and circumstances would separate us, but our special bond lasted. Linda and I agreed that we would return again to Hawaii and see more than just Honolulu and visit more authentic Hawaiian places.

It was a successful training period. Favorable performance evaluations showed that our reservists were productive and appreciated. I received an excellent fitness report from Captain Kelley, the active-duty commanding officer at the Submarine Base. Things were looking up.

What did I learn?

✓ With over fifteen years of naval service, assigned to several ships, schools and shore commands, I had encountered and experienced many leadership styles, most of them were sound. Some were not. Dressing down, berating and outright screaming at subordinates created negative emotions, resentment and reduced morale. Compliance and performance were short lived. Unfortunately this was common in the 1960s.

- ✓ The best leadership styles used coaching, teaching, mentoring and motivational approaches. Praise in public and discipline in private. One of the most amazing and uncommon was the ability for some to "think outside the box," to see possibilities and opportunities when others saw only difficulties and distress.
- ✓ One of the best examples of this was Admiral Nimitz. I believe the story is mostly accurate. No. I didn't serve with him. I just read about him. He was a qualified submarine officer, appointed by President Roosevelt as commander in chief of the US Pacific Fleet following the December 7th Japanese attack on Pearl Harbor. He flew to Hawaii and toured the harbor on Christmas Day 1941. The boat driver, known as a coxswain in the Navy, asked the admiral what he was thinking. Nimitz said (in essence) that the Japanese had screwed up. "We can still fuel our fleet." The Japanese did not bomb the oil storage facilities or destroy the dry docks. "We can repair our ships." They attacked on a Sunday when thousands of our sailors were ashore. "We can man our ships."[14] While most saw only destruction and death, Nimitz saw the seeds of future naval victory in the Pacific.

---

14 "Reflections on Pearl Harbor" by Admiral Chester Nimitz. A Day that will Live in History. Seeking Alpha. December 7, 2011. https://seekingalpha.com/instablog/388783-christopher-menkin/242946-reflections-on-pearl-harbor-by-admiral-chester-nimitz

- ✓ I would add that the Japanese did not attack the Pearl Harbor Submarine Base or any of our in-port submarines.
- ✓ Workers at SubBase Pearl Harbor went on to prepare and repair our fleet submarines for war patrols in the Pacific. Their skills and labor helped win the naval war in the Pacific.

CHAPTER 29

COMMANDING OFFICER

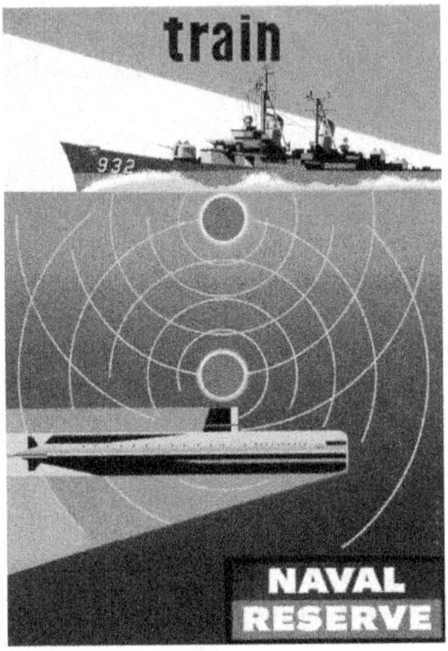

Painting by Joseph Binder, C. 1960, courtesy NHHC, Accession #68-084-A-12.

When I rejoined the Naval Reserve in 1974, after ten years of active duty, my goal was to make lieutenant commander and retire in 1981 with twenty years of service. My thought was, serving and advancing in the Twin Cities Submarine Base Pearl Harbor Reserve Detachment would be a good place to finish up. It was a good Navy home with outstanding fellow officers and sailors, and I was eager to stay there. However, in the military, there are two, often-competing demands: 1) the needs of the individual and 2) the needs of the service. And the needs of the service always win.

On a drill weekend in the summer of 1978, I was summoned to meet with Captain John Francis. He carried the same name as my grandfather, but we were not related. Captain Francis was commander of Naval Reserve Readiness Command Region Sixteen (REDCOM16).

He was the officer in charge of all of the Naval Reserve centers in several Midwestern states. When I received the "invitation," I thought back to nearly a decade earlier when I was invited to meet with the professor of Naval Science at the University of Louisville, who asked me if I would like to become a Navy pilot. Why would REDCOM16 want to see me, a young Lieutenant in Minneapolis?

Captain Francis got right to the point. He said, "Lieutenant, there is a Submarine Base Pearl Harbor Reserve Detachment in La Crosse, Wisconsin, that cannot pass an inspection. They have failed their last two readiness evaluations."

I replied, "I'm sorry to hear that, Sir."

"Mr. Francis, I want you to go down there as commanding officer and turn the unit around."

Rather than feeling flattered, I felt annoyed. "Well sir, I'm really happy here in my current assignment."

"Lieutenant, the first time I discuss this with you is a request."

"I see, Sir. How far is La Crosse from here?

The captain said, "About 120 miles. I'll have my staff draft your orders."

On October 1, 1978, as ordered, I assumed command of Submarine Base Pearl Harbor Detachment 1016 in La Crosse, Wisconsin, a twenty-five-member unit. But I came in mad and stayed mad for several months. I didn't want to be there. I regarded it as a hardship rather than an opportunity. The assignment required that I drive hundreds of miles, pay for my own travel, lodging, and meals, and be away from home from Friday night to late Sunday night. Consequently, I approached the assignment ungratefully with a noticeably bad attitude.

When watching the Green Bay Packers game on Sunday afternoons was in the Plan of the Day, I knew it would be a challenge. Sadly, it was a laid back, good-ole-boy culture. Assigned to me were two junior lieutenants, whom I regarded warily, wondering why they had not been selected for commanding officer of the unit.

My unhappiness and irritation were apparent and resulted in some honest and fruitful conversations with my fellow officers. What do we need to do? We agreed to reorganize the Wisconsin unit similar to the larger Minnesota unit I had just left. Frequently, I shared (more likely berated them about) the rapid success we had achieved in the Twin Cities.

We began to build a leadership team that included two chief petty officers who were eager to lead a successful organization. We implemented the blueprint from my previous organization with the time-tested and excellent training schedule. Next we found volunteer instructors. I relentlessly practiced a top-down, directive management approach, and implemented a management-by-objectives system.

It's a long naval tradition that the executive officer is the disciplinarian, and the commanding officer is the inspirational leader. We reversed that. Jack, my executive officer, was approachable and amiable. I was neither. I was stern and demanding. My "take no prisoners" attitude led to several enlisted men's retirements.

Fortunately, my first reserve unit command was saved by two-dozen dedicated and determined Midwestern sailors who responded positively and coalesced into a highly functional and disciplined unit.

The following year, we passed our Command Readiness Examination. Unit cohesion was bolstered in April during our first Active Duty for Training at the Submarine Base in Pearl Harbor, which raised everyone's morale and sense of unity.

Polynesia

My entire family, Linda, Robin, Jocelyn, and Melissa went to Hawaii with me that April. We stayed at the Hale Koa. On this trip, Linda was determined to get us out of Honolulu. First she took us to the Polynesian Cultural Center founded by the Mormon Church to preserve Hawaii's Polynesian roots.

It is an authentic Hawaiian village that offers ukulele, weaving, cooking, and hula dancing lessons. Our three daughters took hula dancing lessons. Talented villagers share the beauty and culture of the island where sky, land, and sea work in harmony. *He la nani keia la* (today is a beautiful day.)

Next we circled the island to visit Oahu's main areas: Honolulu, the North Shore, the Windward Coast, Central Oahu, and the Leeward Coast. We crossed Oahu's two mountain ranges.

Our family snorkeled at the beautiful Hanauma Bay Nature Preserve, viewing a variety of colorful, tropical fish. On Waikiki, after breakfast each morning, the girls needed to decide whether they would go to the beach or to the pool first, just like our time in Virginia Beach, Virginia. Linda made our time in Hawaii active and memorable.

Linda and Jack, my executive officer, organized a La Crosse unit picnic on the grounds of Fort DeRussy Military Reservation Park, a unique area of preserved green space on Waikiki Beach next to the Hale Koa. As I was intent on learning how to cut a pineapple, Linda noticed a hungry and homeless man near us. She mentioned that to Jack and they both compassionately prepared and delivered a plate of food to him.

While I was on base, Linda and our daughters respectfully visited war memorials: the sunken Battleship Arizona Memorial, and the Punch Bowl National Memorial Cemetery of the Pacific. On a workday, on my lunch hour, in uniform, I left the Submarine Base to visit the USS *Bowfin* Memorial, a preserved World War II, diesel-electric fleet submarine moored in Pearl Harbor. This submarine, along with others, sank tons of Japanese ships. Ironically, *Bowfin,* that afternoon, was filled with Japanese tourists.

As I walked through the familiar boat, I was stopped by an elderly Japanese woman. Seeing an American naval officer in uniform, she asked if she could take my picture.

"Certainly" I said, and stood at attention.

She said, "No. Do something. Do something."

I reached up and put my hand on the nearest valve. She took my picture, bowed courteously, and thanked me.

Returning to the Reserve Center in La Crosse, I experienced a massive release of pressure and anxiety. My respect and affection had grown for those sailors who had tolerated my autocratic approach and carried us over the finish line.

A year into my unwanted assignment, I began to enjoy the job and loosened up. At the end of that year, my senior chief took me aside and said, "Captain, when you first arrived we weren't sure you were human."

What did I learn?

- ✓ Some assignments are harder than others. Maintain a sense of service and a positive attitude in the face of difficult duties.
- ✓ Management by objectives (MBOs) is a time-tested method of aligning management and employee objectives. Involving the team in goal-setting and work-planning promotes buy-in and commitment. The term was first outlined by management guru, Peter Drucker, in his 1954 book, *The Practice of Management*.
- ✓ If you fail to plan. You plan to fail.
- ✓ Again. Ship (Military Unit). Shipmates. Self. In that order.

CHAPTER 30

LCDR

Image courtesy of military-ranks.org.

The La Crosse Submarine Base Pearl Harbor sailors pulled together, rowed the boat, passed our Readiness Inspection with a high grade, and helped me to get promoted to lieutenant commander. In gratitude, I prepared letters requesting the Navy Achievement Medal be awarded to my leadership team. Even if it was not approved by "higher authority," I wanted to publicly recognize them for their efforts and achievements. "Success has many fathers."

Happily, they received the award, "For superior achievement in performance of your duties . . ."

Year two in my tour as commanding officer was dedicated to refining our organizational structure, training, and record keeping. *No Records = No Achievement*. The La Crosse Reserve Center commanding officer informed me that I could do a better job communicating up the chain of command and sharing our training resources with other reserve units in La Crosse. My intense focus had been down the chain of command. Yet again. Once again. I was not adequately managing upward.

I agreed to be a better communicator and "cooperator."

It was a fiscal year with tight Naval Reserve travel budgets. Our unit would not be going to Hawaii. Not enough money. Air fares to Hawaii were pricey. Pick another place. My proposal was that my unit serve aboard a submarine tender in San Diego. San Diego, California, is the principal base for the US Pacific Fleet with over forty-six Navy ships and 213 commands.

Submarine tenders have the same shops and the same mission as a submarine base. They repair submarines. Our request was approved along with a cost-saving recommendation that we hold our drill weekend in San Diego ahead of our active-duty time--fourteen days instead of twelve. Sounded like a great plan. "All battle plans are good until you meet the enemy."

We landed in San Diego for a weekend at the San Diego Reserve Center to discover that the San Diego Reserve Center was not open that weekend. My training team (and me) had failed to check on that detail. Everyone turned to me for a new plan. When you're the CO, that happens.

At the time there were only males in the reserve unit, and I would often address them as "men." I directed the men to check into the barracks at the San Diego Naval Base for the evening. Then muster in the morning for the Plan of the Day, whatever that might be.

Armed with maps of the base, we found the Enlisted Berthing and the BOQ (still called Bachelor Officers Quarters.) In waning daylight, I went down to the waterfront to check on possible activities. What I found were several small Navy sailboats, and I booked all of them for the next day. We would conduct seamanship training.

California Sailin'

My chief in charge was in charge of marching the unit, in the morning after breakfast, to the waterfront area I had marked on the map. We often marched in the drill hall in La Crosse to get out of the classroom and stretch our legs. SubBase La Crosse Detachment 1016 sailors were good at marching.

Using the sailing knowledge I had gained at Officer Candidate School in Newport, Rhode Island, in the summer of '69, I instructed each sailor on hoisting, setting, and dropping the main sail. There was only one sail. I provided further instructions on the components of the sailboat, the operation of the tiller, mainsheet, and how to read the wind using the telltales.

The young, enlisted woman in charge at the harbor allowed the lieutenant commander/ sailboat captain, who appeared to know what he was doing, take charge of the operation. Taking the lead boat, all followed me. When I shouted a command like "ready about" or "hard alee," it was echoed by each sailor in the formation behind me. It was a testament to their Navy training and discipline.

Together we went through beating, reaching, and running, tacking port and starboard. There were no mishaps. We circled in one direction, then another. Their novice skills were impressive.

Luckily, the winds were light. As we headed back into the docks, I dropped the main and eased into a slip. They all followed. No one crashed into a pier. After lunch we set out again.

This time I informed them that we would do figure eights, like a ship does in a man-overboard drill or actual situation. "There will be risk of collision, so be alert."

Everyone was finding this activity in the warm Southern California sun very enjoyable. The Wisconsin and some Minnesota sailors were fast learners. We did several figure eights with no collisions. Thinking that was enough risky behavior for the day, I led them back to the landing. However, I offered my students one last challenge. Bringing my boat in, I turned into the wind, pulled the main sheet in flat, and coasted into the dock. They all tried to copy the difficult maneuver. There were many minor bumps.

I complimented two dozen smiling sailors, some with sunburns, on their new seamanship skills. We were putting the Navy's limited training funds to good use.

"Have a great evening. Tomorrow, I've arranged damage control training in the wet trainer. Wear your dungarees."

The Resolve Academy wet trainer welcome: "IMMERSE YOURSELF IN OUR 2-STORY TRAINING FACILITY WHERE FLOODING, BREACHES, BROKEN PIPES AND HATCH FAILURES ARE SIMULATED USING REALISTIC FLOOD CONDITIONS. Teams work together making the right decisions, activating control tactics, and developing hands-on skills in a safe but intense environment."[15]

The next morning, we mustered at the wet trainer. After lengthy instructions, we entered the vessel, and I joined the crew.

My leading chief said, "Captain, are you sure?"

"Yes, Chief. But you're in charge of the damage control team. And please don't let the water get above my chin."

As the water rushed in from several openings and began to rise in the compartment, I watched "my men" work quickly and smartly to stop as many leaks as they could. They worked well together, using the damage control materials available to them, while maintaining discipline and a can-do attitude. When the operators finally shut off the flooding, the cold water was up to my chest. We all looked at each other, soaked, cold, and full of pride.

Many of the men said to me afterward, "Captain, that was the best training and the most fun I've ever had in the reserves."

Again, we were good stewards of our travel dollars. We got lemons and made lemonade.

Monday morning, as ordered, we reported aboard the USS *Dixon* AS-37 for our two-weeks active duty for training. USS *Dixon* was the second ship in the L.Y. Spear class of submarine tenders designed and fitted to accommodate attack submarines and service up to four submarines moored alongside simultaneously. Dixon won numerous awards for technical and service excellence. However, as a Northern Yankee, I thought it was unfortunate that

---

15 Resolve Academy. https://resolveacademy.com/facility/damage-control-wet-trainer

Dixon was named in honor of George E. Dixon, commander of the confederate submarine *H.L. Hunley.*

USS *Dixon* in San Diego, CA.
courtesy NHHC, catalog #NH 107769-KN.

We were all assigned to the large repair department. The enlisted members reported to their appropriate shops. My assignment was assistant repair officer. Dixon's repair officer, a commander, gave me a tough job. Find out what's wrong with the electric motor in one of the "B Girls." The B Girls, as they were known in the submarine community, *Barbel, Blueback, and Bonefish*, were the last diesel-electric submarines built by the US Navy. They incorporated a number of experimental features that were tested earlier by the USS *Albacore.*

USS *Barbel* SS-580 launch 19 July 1955, courtesy NHHC #UA 468.06.01.

Among the prominent features were the streamlined, teardrop, double-hull design, and placing the conning tower, attack center, and control room in the same space.

With supreme confidence, I accepted the assignment and believed that collaborating with the electricians on board the submarine and the tender, manufacturer's technical representatives and Naval Sea Systems Command (NAVSEA), we would figure it out. Never having been aboard one of the B Girls, I found the boat intriguing, seeing for the first time how they were built and operated.

However, twelve days later, I departed *Dixon* disappointed that we had not solved the mystery. But I did leave behind a detailed record of our troubleshooting, which yielded a good fitness report for my "valiant" efforts.

What did I learn?

- ✓ Naval officers (and all military leaders) lead and serve because of a sense of duty, mission, patriotism, care for those in our care, professional pride, and to get a good fitness report for retention and promotion.
- ✓ Always affirm, recognize, and reward accomplishments and achievements.
- ✓ Public praise motivates good behavior and good performance.
- ✓ The Midwestern United States grows good sailors.

CHAPTER 31
BRINGING JON HOME

"And it came to pass . . . that the soul of Jonathan was knit with the soul of David and Jonathan loved him as his own soul."
—1 Samuel, 18 (NRSV)

The United States was in our fifth decade of the Cold War with the Soviet Union, and our Federal Government was pairing tax cuts with massive increases in defense spending. Trickledown economics even trickled down to the Naval Reserve. With significant increases in the US military defense budget, designed to bankrupt the Soviet Union, came increased travel dollars for the Naval Reserve. My SubBase La Crosse Unit would go back to Pearl Harbor. At my request, I was extended for a third year as commanding officer. My reserve unit had reached a level of sustained excellence in training and administration, peak readiness, and harmony in personnel relationships. We all reaped the results of our earlier stress and labors.

By now I had twenty years of naval service, had been promoted to lieutenant commander and qualified for retirement with pension (at age sixty for a reservist). However, I decided to stay in. My next goal was commander with twenty-six years of honorable service, which would improve my pension. It made financial sense.

Myrtle Francis

On New Year's Day, 1982, my ninety-two-year-old grandmother died. She was my father's mother, and she lived with us for a few years. She was loving, kind, and affirming. I was indebted to her for many reasons. In an unselfish act of kindness, she took all of my new Navy uniforms after I enlisted and sewed my one seaman recruit stripe on each jumper and shortened my bell bottom trousers.

She sang as she did it. It is one of my fondest memories. I found out years later that she was married at a lumber camp in Michigan when she was seventeen. She married John Francis and later gave birth to my father. She was the best grandmother on the planet.

At the end of my tour of duty as the commanding officer of the Submarine Base Pearl Harbor Reserve Unit in La Crosse, I received an excellent fitness report from Captain John Francis (still no relation), the Readiness Region commander, who had sent me to La Crosse three years earlier.

It was a sad farewell. We had bonded, the La Crosse sailors and me, a grumpy, demanding outsider from St. Paul. But I would not miss the travel and the expenses that came out of my own pocket. However, I was out of a job. After years of absence, I was out of the loop in the Twin Cities Naval Reserve Center. Working with the Reserve Center commanding officer, I applied for several open billets.

I landed a new assignment as training officer of the Personnel Mobilization Team. Their mission was to advise other units on their readiness for mobilization and acceptance at their gaining commands. It was an easy transition back to third officer, responsible only for training. My reserve unit training toolkit was full of tools. However, I did have some new learning to do.

I scheduled an active duty at Naval Reserve Headquarters, The Naval Reserve Personnel Center in New Orleans, Louisiana, where I would study policies, procedures, functions, and administration of the Naval Reserve. But I made a mistake by going to New Orleans in January.

Expecting warmer, Gulf Coast weather, I packed light with only short-sleeve khakis and a raincoat. It was unusually cold in New Orleans that month. One evening it went down to 17 degrees Fahrenheit. I struggled to keep warm.

The water pipes froze because most of them were above ground. For one day we had no running water. My officer accommodations were located across the Mississippi River from Naval Reserve Headquarters. Each day I rode a small Navy utility boat across the river. Always being the senior officer aboard, I was in charge of boat safety "in accordance with" long naval tradition. My knowledge of the wide river, the tides, or the heavy shipping traffic was negligible, yet I was in charge.

Many mornings, in the cold weather, the air was colder than the water, and we drove through thick fog that blanketed the big river. That made the crossing dangerous because there is a lot of ship traffic on the river around New Orleans.

As I boarded the utility boat, I greeted the Coxswain, and asked, "Have you done this before?"

Each day his response was the same, "Yes sir, we will be fine."

We safely made the crossing twice a day for ten days.

A call from Linda requested that I go to the Bombay Company in New Orleans and buy several pieces of furniture. I packed the furniture securely for my flight home. After my training was complete, I went to the airport, checked in my seabag and several boxes of furniture, and we lifted off for Minneapolis.

On time, we arrived outside of the Twin Cities airport when the captain announced that, due to a major snowstorm, the airport had just closed. We circled back and landed in Memphis, Tennessee. After two weeks away, I was anxious to get back to my family, warm fires, and running water, so I took the next flight out of Memphis when the Twin Cities airport reopened.

We landed at midnight. Together with some fellow passengers, I rented a limousine. There was twenty-two inches of snow on the ground when I arrived at my house. I dragged my seabag and furniture through waist deep snowdrifts and announced at 1:00 a.m., "Hi, honey, I'm home."

Linda, pregnant with our fourth child, had waited up for me.

When Linda announced to our family that she was pregnant, at thirty-seven years old, our youngest daughter began to cry. Melissa was afraid that her mother would not survive childbirth at such an advanced age.

Jonathan David Francis

Jonathan is Hebrew for "God's gift" and David means "beloved." We had our son's name picked out for twenty-two years. When Linda and I were high school sweethearts, we talked about a time when we would be married, have children together, and raise a family.

On March 5, 1982, someone I had waited for my entire life finally arrived. He had my dark-brown hair, a handsome face, and sincere brown eyes that I thought reflected his Irish heritage.

Our son made me feel whole. Our family was now fully formed. We had three beautiful and healthy daughters, ages eleven, fourteen, and sixteen, and the son Linda and I had held in our hearts for twenty-two years.

When our daughters were babies, I was on active duty, going to sea as a submarine sailor and later as an officer. Linda and I had three children in five years, but I was seldom home to help with our girls. I missed so much of the joy *and* the hard work of their early months.

Jonathan was my first chance to be present during the baby months. I cherished that time—the baths, even the diaper changes and the night-time feedings. I rocked him and fed him his supplemental bottle, relishing our time together in the quiet beauty of the night.

Once again, 1982 was a year when we felt the sorrow of death and the joy of new life. In a few years, we had a child in college (Robin was attending the University of Massachusetts - Amherst) and a child in diapers (Jonathan was snuggled in a crib in the nursery).

What did I learn?

- ✓ Never miss an opportunity to tell your loved ones that you love them.
- ✓ Check the weather forecast at your destination before you leave home.

CHAPTER 32

TRIDENT

Trident Missile launch from USS *Michigan* SSBN-727, courtesy NHHC, catalog #L55-19.06.06.

Naval Base Kitsap (NBK) is located in Washington State. Bangor, Washington, was selected as the homeport for the first squadron of the Ohio-class of Trident ballistic missile submarines and the Strategic Weapons Facility Pacific. The facility maintains and stores Trident ballistic missiles.

Poseidon was one of the most powerful gods in Greek mythology. The Trident is the magical and powerful weapon of Poseidon, god of the sea. So naturally, following the Poseidon missile, the US Navy named its next submarine launched ballistic missile, Trident.

After my required trip to Naval Reserve Headquarters in New Orleans, I was free to select my next active duty for training site. Our West Coast Trident Submarine Base is on Washington State's Kitsap Peninsula. I wanted to get out there, see it, and get up to date as much as possible in my warfare specialty: submarine warfare.

Submarine Base Bangor was one of our newest. It was so attractive and well-provisioned, that our sailors said it's "as fine as an Air Force Base." The US Air Force, having been established shortly after WWII, had newer military bases than the US Navy, which was established in 1794. Why a West Coast base? Because of the nuclear threat posed by China, North Korea, and the Russian Federation.

Our Trident armed submarines were now the US Navy's newest and largest fleet ballistic missile submarine. The US built eighteen Trident submarines, named for states in the union, with *Ohio* as the lead ship in the class. Combined they carry approximately 50 percent of the total US inventory of strategic thermonuclear warheads.

The US military-industrial complex went big on this one. The size and scale of this submarine weapons system are enormous: Trident subs are 560 feet long, 18,750 tons, with a draft of 35.5 feet. They displace more water than US Navy cruisers and carry twenty-four missiles. The Trident missile weighs an incredible 130,000 pounds and is forty-five feet tall.

But they were not the largest submarines in the world. Our Trident subs were in third place. The largest submarines are the

Russian Navy's Soviet-designed 48,000-ton Typhoon class and the 24,000-ton Borei class.[16]

In June I spent fourteen days at Submarine Group Nine in Bangor. It rained for thirteen of those days. The Submarine Group oversees all of the Bangor Submarine Base. Assigned to the Training Department, I was given a project to improve Trident communications. Doing well on the project, I was asked to return the following year with the advice that I come back in August, when the weather is sunnier.

Returning to Bangor, the following year in August as advised, I was assigned to Commander Submarine Squadron 17, The Trident Squadron responsible for maintenance and repair of our Trident submarines. Assigned an office at Delta Pier with a stunning view of the Olympic Mountains, I worked on several challenging special projects that got me aboard two Trident submarines: The USS *Michigan* in drydock, and then underway on the USS *Florida* for operations in the Hood Canal.

---

[16] Lockheed Martin. Trident ii d5 Fleet Ballistic Missile. Overview. 2021. https://www.lockheedmartin.com/en-us/products/trident-ii-d5-fleet-ballistic-missile.html

USS *Michigan* SSBN 727, courtesy NHHC.

Florida was similar to *Lewis and Clark* in layout and equipment, only much larger. On board *Florida,* I walked to exhaustion, inspecting her systems and spaces. Since I had not been aboard a submarine at sea for many years, for lunch in the wardroom I instinctively took a seat at the junior end of the table. The captain sitting at the head of the table motioned to me to sit next to him. "Commander, sit up here."

My tours of duty at the Trident Base were professionally rewarding and informative and recognized with excellent fitness reports. My reserve officer career was going much better than my active-duty officer career.

What did I learn?

- ✓ Never stop learning.
- ✓ Lifelong learning can be good for your career and for your life.
- ✓ Knowledge is power, and it can also be life-altering, useful, interesting, and stimulating.

CHAPTER 33
COMMANDER

Image courtesy of military-ranks.org.

Commander is the first senior commissioned officer rank in the US Navy, equivalent to the rank of lieutenant colonel in the other Armed Services. Commanders wear the navy-blue uniform of a full line officer and are the first rank permitted to wear the gold-embellished cap brim, often called "scrambled eggs."

Image courtesy of military-ranks.org.

From white hat to senior officer "headgear" in only twenty-five years! The first time I wore my hat with the gold-embellished brim, I felt strange. It took me several months to feel authentically like a senior naval officer.

Admiral's Staff

One of the most coveted positions in the Naval Reserve for an officer is assistant program officer at the Readiness Command. My mentor, Commander Anderson, had served there and recommended that I apply.

While a lieutenant commander, my application was approved for a position on a Naval Reserve admiral's staff, and I reported to the inspector general. We staff officers conducted readiness evaluations at numerous Midwest reserve centers throughout Readiness Command Region 16.

My years of experience being on the receiving end of inspections, preparing for them, organizing units, and constructing excellent training programs, prepared me well for this new role. It was a traveling job, but this time the Navy covered the travel costs.

It was fun and enlightening to be an inspecting officer. My approach was that of a consultant. I conducted an analysis of the reserve unit's organizational health, problems, and achievements, and developed a plan with recommended improvements in their training, administration, and readiness. Unfortunately, I did have to give a failing grade to some units where it was obvious they were not even trying.

At the end of this rewarding two-year tour of duty, I returned to the Twin Cities Reserve Center. My growing service jacket now contained an excellent fitness report, from a Navy admiral, with his recommendation for early selection to commander.

And what do you know—the following year, I was selected for US Navy commander!

After departing the Readiness Command staff, I was directed to a reserve unit that supported fleet repair services to the Pacific Fleet. They needed an executive officer. When I met with the CO of the unit, Commander Dave Recker, I could feel his energy and enthusiasm.

The fifty-member reserve unit was attached to the USS *Cape Cod* AD 43, a destroyer tender in San Diego. Training officer three-times, commanding officer for three years, but I had never been an executive officer (XO). The XO is the human resources executive, responsible for personnel and administrative management. I knew the job description, and I had supervised one and evaluated several executive officers. This was my chance to serve in the position.

The CO, Dave Recker, and I had both attended the Navy's Leadership Management course. It is the best management training I ever received. The most important takeaway for me was the "situational" leadership concept. People are unique and respond to different styles and degrees of leadership. Some require close supervision while others just need to be given broad goals. Cookie-cutter leadership approaches are not the most effective. If only I had attended the course before I became commanding officer in

La Crosse. Dave gave me broad goals and empowered me to take management actions to meet those goals.

This was my first coed unit. There were several female petty officers in the Cape Cod Reserve Detachment. Prior to this assignment, I had served in the US Navy for twenty-four years in male-only ships, stations, or reserve detachments. However, most reserve officers like me had extensive experience collaborating with women in the corporate world.

Nevertheless, I became hypervigilant to how females were treated and stressed equal opportunity programs, courtesy, and consideration. Constantly alert for possible sexual harassment and gender discrimination in assignments, I determined that it would not happen on my watch. Navy women, enlisted and officers, would soon serve in every specialty aboard Navy ships, including surface warships and submarines.

When I joined the Cape Cod unit it had an R-3 readiness score. That's low. Our goal was to get to R-1: Fully Ready to Deploy. Learning from my past experience, I implemented a management by objectives (MBO) system and put senior enlisted members in leadership positions. It was the early stages of personal computers (PCs) and I was able to load the information on my own PC and print out and distribute our management reports.

My leadership toolkit now contained two new tools: situational leadership and electronic management reports.

The new leadership team and I were tracking progress in all areas: unit training objectives, schedules, class curriculum, personal rate training and advancement, active duty for training, and administrative requirements. Similar to my experience at 3M, and Dave's experience at The Toro Company, the management reports were reviewed, studied, and updated, and appropriate corrective action was taken.

Within nine months, our Cape Cod detachment attained an R-1 readiness score. I had learned in La Crosse that *No Records = No Achievement*. We had a massive supply of pertinent records, and they were now stored on a hard drive.

USS *Cape Cod* AD 43 was not available when I attempted to schedule my active duty for training for June. However, I received orders to a sister ship the USS *Acadia* AD 42. This was my first tour on a surface ship tender, but I suspected she was similar in organization to *Dixon,* the submarine tender on which I had served.

Destroyer tenders (ADs) are named after geographic locations in the United States, hence Cape Cod (The Massachusetts cape extending into the Atlantic Ocean) and Acadia (the maritime region that includes the US State of Maine.)

Reporting aboard *Acadia* in San Diego, I was assigned as an assistant repair officer and handed some challenging assignments: improving personal radiation exposure procedures, and planner and team leader on a Battle Readiness competition. Unlike the frustrating and unresolved submarine electric motor problem I worked on aboard *Dixon,* I was able to successfully complete these projects.

One afternoon, I took lunch in *Acadia's* junior officers' mess. It was quicker and less formal. Across the table from me were a warrant officer and a female ensign. She was *Acadia's* training officer, and she was having a bad day. Unfortunately, she chose to complain to the warrant. She couldn't see, but I could detect his growing annoyance.

Suddenly he turned to her and said, "Do I look like the duty rag holder?"

She began to cry.

I reprimanded him. Mumbling under his breathe, he stood up and left. At that time, my Navy was not doing a good job of integrating woman into its "ancient seagoing fraternity," and I felt a deep sense of disappointment.

At this writing, the US Navy has significantly improved its integration of women aboard ships and stations. We've made great strides. The first American woman currently scheduled to step on to the moon is a female US Navy lieutenant commander, Naval Academy graduate, and qualified submarine officer.

Acadia actually got underway. We were bound for San Francisco. When we departed San Diego, I thought of the old joke: Tenders are sent to sea so the base can dredge the coffee grounds up alongside the pier. Entering San Francisco Bay, passing under the Golden Gate Bridge was a once-in-a-lifetime thrill.

At the end of my time on *Acadia,* I received an excellent fitness report from the ship's captain. Our United States Surface Navy is a powerful force to preserve the peace and project US Naval power. But I was anxious to get back to the submarine service. When I learned that the Submarine Base Pearl Harbor detachment commanding officer billet in the Twin Cities would be open, I applied for it. It was a commander's billet. And I was now a commander.

Commander Recker wrote a superb letter of recommendation for me and I got the commanding officer assignment. I was heading back to submarine world.

What did I learn?

- ✓ My US Navy did not do well in the early years integrating women into its ranks. However, we seem to be making progress.
- ✓ "We cannot all succeed when half of us are held back." —Malala Yousafzai.
- ✓ "As a Navy, we celebrate the women who have and continue to serve in our ranks. The numerous accomplishments women have achieved in the Navy were made through hard work and grit. Their perseverance is as much of an inspiration as is their continuing achievements." — Chief of Naval Personnel Public Affairs. - MCI Mark D. Farem, August 26, 2017.

---

17 "Celebrating Navy Women, Perseverance and Achievements." MCI Mark D. Farem, https://www.navy.mil/Press-Office/News-Stories/Article/2325218/celebrating.

CHAPTER 34
SECOND COMMAND

Members of the Twin Cities Reserve Detachment in Pearl Harbor, Hawaii. (Commander Francis, far left, on USS Los Angeles gangway.)

It felt like a homecoming and a return to a familiar community when I reported as commanding officer of the seventy-member, Twin Cities Submarine Base Pearl Harbor detachment. Many from the original crew, dating back to 1977, were still there. The detachment had added a handful of female petty officers, and the unit was in good shape.

It had been almost ten years since I served as its first training officer, yet many of the training programs and the basic organization I had introduced were still in use. My goal would be to maintain good performance but look for areas to improve. "Avoid complacency and always be on a path of continuous improvement." I found one: the unit had dropped below the top R-1 readiness score. That gave me something to work on.

Since I was an experienced inspector, I decided to hold an inspection on my new unit. What I found was previous officers had not kept pace with or up to date with changing readiness metrics. To meet the new requirements, I provided a roadmap to the unit leadership. We got back to R-1 in nine months.

Months later, I was informed at the Submarine Base Pearl Harbor Commanding Officers' Conference in San Diego that my detachment was the only Pearl Harbor unit at the R-1 level. But my sense of achievement was dampened when a new Naval Reserve organizational plan arrived. My detachment would be combined with two others to form a larger 225-member Readiness Unit, and I would be its executive officer. Many SubBase sailors were highly upset and wanted me to stop the transition. But I had no ability to do that. It was like a hostile takeover in business with a risk that we might loss our unit cohesion and high morale that was a hallmark and source of pride.

At the SubBase Pearl Harbor Commanding Officers' Conference, I was asked to speak to the gathering and express my opinion. In my brief talk, I said that the reorganization may be good for the Naval Reserve, but may not be beneficial for the Submarine Base in Pearl Harbor. There was risk of readiness and morale decline. The admiral present at the conference took me aside afterward and said, "Commander, that took a lot of courage."

I thanked him, thinking, *I may have just screwed up (again).*

Commanding Officer's Conference

(Commander Francis, far right in front row)

What was once, a short time ago, an enviable position as commanding officer of a small, exceptional organization turned into one of my most difficult management assignments: merging three mobilization detachments into one—when one of the units didn't want any part of it. It was a tough sell. And it was a tough time. Our remaining parents died one year apart. First Keith, Linda's dad in 1987, then my mother, Millie, in 1988.

Our family traveled back to Michigan for each funeral. Robin flew in from college in Massachusetts. A young boy, Jon showed his compassion and maturity through each service. It was the saddest of times. My mother's death hit me hard, and I was in deep grief for many weeks. Linda and I felt like orphans. Neither of us had parents. And our children no longer had grandparents.

On November 9, 1989, crowds of people in Germany began to dismantle the Berlin Wall. The wall had been a vivid symbol of the Cold War that had divided Europe for close to thirty years. The exciting news that the wall was coming down brought back memories from my Navy boot camp. I was a seventeen-year-old seaman recruit when the wall was constructed; and now, twenty-eight years later, I was a Navy commander.

I could see that throughout this long Cold War, I had made personal progress, and a small contribution to this victory. The removal of the wall was a symbol of the end of the Cold War—a national achievement. We prevailed—that felt good.

What did I learn?

✓ The only constant is constant change.

✓ "One can choose to go back toward safety or forward toward growth. Growth must be chosen again and again: fear must be overcome again and again." —Abraham Maslow

✓ "The secret of change is to focus all of your energy, not on fighting the old, but on building the new." —Socrates.

✓ "If we love, we grieve. Grief is the bleeding heart of love." —David Francis, *Grief Travelers*.

CHAPTER 35
RUXO

In my new, expanded reserve assignment in 1989, I adopted the abbreviation RUXO, Readiness Unit executive officer, and inherited eight officers and 225 sailors from diverse backgrounds. Working with Captain Jim Erickson, the commanding officer, we agreed on an organizational structure. The result resembled General Motors, with a corporate entity on top of three somewhat autonomous units headed by their former commanding officers. That meant I would wear two hats: Readiness Unit XO and SubBase Pearl CO. That was the easy part.

The hard part was to convince SubBase Pearl Harbor leaders to join the corporate entity and provide consolidated training and management. Many rose to the occasion. *In Search of Excellence, Lessons from America's Best-Run Companies*, by Peters and Waterman, Jr. guided my work. We created a loose/tight organizational structure.

The three units had some flexibility to meet their unique gaining command requirements, but reported to the "head shed" for overall readiness attainment, including training and personnel performance and unit evaluations. My pitch to SubBase Pearl Harbor was an opportunity to build something new, learn expanded leadership, while at the same time maintain unit cohesion. And after all, we still get to go to Hawaii once a year.

The Readiness Unit start-up went well because our senior enlisted, and most officers, provided sound leadership. Reaching into my old management-by-objectives toolbox, I implemented electronic reporting and pushed responsibility and authority as far down the chain of command as possible. It worked.

One of the memorable highlights of my tour was the arrival of a veteran warrant officer from California. I interviewed him to determine his strengths, career objectives, and experience and where he would fit best in our organization.

"Why did you move from Oakland, California, to Minnesota?" I asked.

He said he was married to a woman from Minnesota, and she wanted to come back home.

"So what do you think so far?"

The warrant responded, "Let me tell you a story. I went to the grocery store and bought a case of beer. Before opening my car door, I placed the case on the trunk of the car next to mine, forgot about it, and drove away. When I got home I told my wife what I had done."

She said, "Call the store. Someone will have turned it in."

I responded, "What planet are you from?"

She said, "No. Call them."

"I did and was told that someone had turned in my case of beer, and I could pick it up at customer service."

With a smile I said, "Welcome to Minnesota."

During our successful active-duty periods at the Submarine Base in Pearl Harbor, we felt that we were part of a small, exceptional organization again. One year on Oahu, I was told that we were challenging the active-duty submarine base softball team. I pulled rank and said that I would be the starting pitcher. Memories of my softball pitching days in Louisville gave me unwarranted confidence.

On the mound that evening, we fell behind 17–0. Every pitch I threw was hit by the active-duty batter. It was obvious that I didn't have the right stuff anymore. But the petty officer/team manager, who organized the game, was reluctant to pull his CO off the mound.

Finally I motioned for him to come out.

He asked, "Are you okay, Skipper?"

"No. You need to pull me."

He did. I thanked him for saving me from more innings of disgrace.

After three years as RUXO, it was time to move on. I requested a transfer and found a billet with the commander in chief, Pacific Fleet detachment. My large organization venture resulted in high praise from my Readiness Unit CO. It was hard earned and rewarding professional praise because he was a by-the-book leader whose management philosophy was "Don't let them see you smile until Christmas."

"CDR Francis is an unsurpassed manager of human resources. His knowledge and understanding of human behavior coupled with his broad mastery of Naval Reserve requirements and directives made him a consummate executive officer . . ."

And I received my last-ever fitness report from an active-duty commanding officer at the submarine base in Pearl Harbor.

As a commander with over twenty-eight years of service, I again considered retiring. It would have been a good time to do that.

Yet in the back of my mind I thought, *Maybe I have a shot at captain. I think I'll stay in a bit longer.*

What did I learn?

- ✓ "Don't reinvent the wheel." (You may have heard this a few times.)
- ✓ There is knowledge to borrow from others.
- ✓ It's out there in books and in people's minds and experience.
- ✓ Find it, ask for it, and use it.
- ✓ Problem solving is a community activity. As a result of future learning, I wish I had taken a more collaborative/problem solving approach in my first CO assignment in La Crosse, Wisconsin.
- ✓ Hold discovery meetings with unit personnel and leaders to uncover reasons why the unit was unable to pass two annual readiness inspections in a row.
- ✓ Assess what's broken. What are the reasons for the failure. What's not working?
- ✓ After the lessons learned are learned, jointly develop a get well plan that addresses the deficiencies.
- ✓ Create a roadmap for success.
- ✓ Develop a classic project plan with action items, timelines and responsible individuals.
- ✓ Share the burden and the responsibility for problem solving, progress and success.

- ✓ A more collaborative approach may have avoided unnecessary stress. I got better at collaborating, but it took me a long time. Among the many things I learned from many people, over 30 years, this is the one that stands out in my memory. I offer it to others.

CHAPTER 36

LAST ASSIGNMENT

Image courtesy NHHC

On August 2, 1990, the Iraqi Army, the fourth largest in the world, invaded its neighbor Kuwait. The United States responded to "preserve the sovereignty of nations" with Operation Desert Shield/Desert Storm. Twenty-one thousand naval reservists were called into active duty. The multinational effort brought together the largest force of Navy warships assembled in a single theater since World War II.

No longer responsible for the "care and feeding" of 225 sailors, I felt relief and gratitude to be back in the warfare game. My new role was expert and trainer in submarine and anti-submarine warfare with the commander in chief Pacific Fleet Detachment. It was a welcome change of pace and an opportunity to go inside Pacific Fleet Headquarters. They just happen to be in Hawaii. On my first active duty there, I stood watch with the operations duty officer.

I was awash in acronyms at a level I had not experienced before. The dense combination of naval, computer, and communications acronyms made reports and messages appear to be written in secret code. Unwisely, I remarked to the active-duty watch officer, "Doesn't anyone write in complete sentences anymore?"

Accustomed to cryptic messages, he viewed my question as strange. CINCPAC was a fascinating place with top-secret information and briefings. I saw the US conduct of the Cold War at a more detailed, secret, and intriguing level.

In addition, I earned my pay and a great fitness report from an active-duty admiral by completing a well-researched special project. But I can't talk about it.

In the fall of 1990, my friend and mentor, Dave Recker, now a captain, reported aboard as commanding officer. I was happy to have the opportunity to serve with him again. We went out on active duty together to Hawaii in February 1991.

Before we flew out, I received a call from my nephew, Navy Chief Petty Officer Daryl (Dale) Tacey, who was stationed in Hawaii. Linda and I had spent enjoyable times with him and his wife Yoshiko on our previous trips.

Dale said, "Hello, Captain."

I replied, "Hi, Chief. I'm still a commander."

He said, "No sir. I'm holding a copy of *Navy Times*. You've been selected for captain."

"Thank you, Chief. That's wonderful! And it's special that I received the news from you."

Operation Desert Shield/Desert Storm was being conducted during my second tour at CINCPACFLT. Not much was going on at the submarine desk, with minimal activity in the Pacific. However, two of our fast-attack submarines launched cruise missiles at land targets (The Tomahawk land-attack missile was first used by submarines in combat during Operation Desert Storm by USS *Louisville* and *USS Pittsburgh.*)

Most submarine support in the Persian Gulf came from the Atlantic side. US submarines protected our carrier groups and ensured that sea lanes remained open. Daily intelligence briefings and message traffic at CINCPACFLT continued to give me a deep understanding of US naval activity, operations, and threats in the Pacific Theater.

Day after day, I read intelligence reports that the Russian Navy was no longer sending their submarines to sea. They remained tied up in port. The Soviet-built 48,000-ton Typhoon ballistic missile submarines were not at sea, prowling off our coasts. After five days of reading these reports, it occurred to me that the Soviets had given up. *We won! We won the Cold War!*

Unfortunately, I couldn't talk about this or share it with my family and friends. Of course, we still faced continued military threats in the Pacific from a more assertive, nuclear-armed Chinese Navy and North Korean nukes. And post-Soviet Russia would continue to behave belligerently. Russia with an economy the size of Texas was still a nuclear power and had an obsession to remain a significant player on the world stage.

The Russian Navy would continue to challenge the US at sea. But our long, costly, and frequently frightening military competition with the Soviet Union was finally over.

Send in the Clowns

On our weekend off, Captain Recker suggested that we tour the Island of Oahu. We drove around the island and stopped at the North Shore to do some body surfing.

Twice in the big waves, I went out and came back in propelled by the surf. It was fun. However, on my third attempt, I was being carried out to sea. The lifeguard stood on his highchair and shouted through his megaphone to me to "come back." But I did not have a megaphone to shout back that I'm being dragged out from shore. And he wasn't coming out to help me.

I abandoned my board and started to swim. But I kept going further out. Resisting the urge to panic, I relaxed, and laid on my back to let the big waves push me in. It was working, and I was heading back to shore! Then a huge wave hit me and pushed me under. I was turning summersaults under water.

While getting dizzy and worried, I began to swallow seawater. For a second time in my life, my life began to pass before my eyes. I was losing hope that I would survive when miraculously the sea spit me out. I landed on the beach feet down and head up. It must have looked like a circus clown performance.

After a brief recovery period, we left for Pearl Harbor. As we pulled away, I noticed a sign that read: Banzai Pipeline. The Pipeline is an area of big surf where surfing competitions are held, where some surfers have died, and where I made the incredibly stupid, boneheaded, mistake to body surf.

For several months, I didn't talk about this near-death experience and my poor judgment. Feeling fortunate to be alive, I looked forward to seeing my family again and being promoted to US Navy captain.

What did I learn (almost too late)?

- ✓ "With heavy walls of water exploding on a shallow and sharp reef, Pipeline and Backdoor are undoubtedly two of the world's biggest waves of consequence. And yet, year in and year out, surfers from around the world travel to Oahu's North Shore and attempt to tame the beast. Some successfully, and some not so much."
- ✓ Captain (Select) Francis is lucky to be alive.

Chapter 37
Captain, USNR (Retired)

Image courtesy of military-ranks.org.

A United States Navy, captain (CAPT), a senior officer rank, has a pay grade of O–6. Above commander and below rear admiral, it is equivalent to the rank of colonel in the other uniformed services. According to Defense Officer Personnel Management Act of 1980 guidelines, no more than 50 percent of eligible commanders should be promoted to captain after serving at least three years at their current rank and 21–23 years of cumulative commissioned service.

You may have noticed a few chapters that began like this with a naval insignia and a description. This is the last one. I did not get promoted to admiral. And I never thought I would ever make captain. But I did. And spanning three decades I did complain and gripe a lot. When I enlisted in the US Navy at age seventeen, and my grandmother sewed my one seaman recruit stripe on my uniforms, I could not imagine that one day, I would retire as a senior naval officer. It did not seem possible or realistic. But it happened. And I could not have done it alone. Many people taught me many things and I have many people to thank:

Linda Francis, my patient, loving, and supportive life partner and Navy wife. She was the glue that kept the family together. She married a sailor and bore the burden of being alone for months, keeping the home fires burning, raising children essentially as a single parent, maintaining our house and supporting my chosen seagoing career.

Robin, Jocelyn, Melissa, and Jon, our children, who forgave my absenteeism, were uprooted too many times, but often felt pride being Navy brats. Expecting a guilt trip, I asked Robin, our firstborn, who moved the most, "What did you think of all the times you had to move?"

She said, "Dad. I actually enjoyed it." Thank you, Robin.

Francis Family on the Beach (1988)
(L-R, Robin, Linda, Melissa, Jon, David and Jocelyn)

My mentors: Captains Al Anderson, Dave Recker, Jim Erickson, and John Francis who brought me along, offered sound advice and gave me opportunities to do well.

Dave Fields and Gordon McGarry, nuclear submarine captains: Thank you for your confidence and for teaching me leadership and seamanship.

Keith Bunting: my wise and compassionate *George Washington* executive officer.

Carol, Carolyn, Linda, Joan, Janet, Diane, Bonnie, Pat, Sue, Tina, and Judy: Neighbors and "Wardroom wives" who cared for Linda and boosted her morale.

Jim and Jerry (WEPS): Thank you for your advice, training, and confidence.

Father Jerry (The Reverend Dr. Jerry C. Doherty), The Reverend Jim Newman, Dr. David Carroll, and my talented editor Marly Cornell: Thank you for your wise guidance and help on my long journey.

Thank you, my fellow reserve and active naval officers who gave me assignments and wrote favorable comments about my job performance.

Thank you to the many sailors I served with on my ships, stations, and reserve units who tolerated me, supported me, and made my many promotions possible.

My fellow citizens and taxpayers who took a chance on a poor kid from Michigan and gave me a path and the ability to serve our nation: Thank You.

Finally, thank you to Navy Chief Daryl (Dale) Tacey. In 2005, I was "called out of retirement" when my nephew Dale extended to me the honor to function as his retiring officer in Oahu. For a few hours, I was senior officer present afloat (SOPA) onboard the USS *Missouri* BB 63, and I was still able to fit into my summer whites. The fantail of the Battleship Missouri Memorial in Pearl Harbor is the site of many Navy retirements.

USS *Missouri* (BB-63) in Pearl Harbor, HI, courtesy NHHC, catalog #DN-SC-92-05364.

The Naval Reserve rarely holds retirement ceremonies. Generally, we all go away quietly, as I did.

In closing, no writing concerning naval matters would be complete without sharing the words of the "Navy Hymn" by William Whiting, 1860.[18]

Eternal Father, strong to save,
Whose arm hath bound the restless wave,
Who bidd'st the mighty ocean deep
Its own appointed limits keep,
O hear us when we cry to thee
For those in peril on the sea!
Bless those who serve beneath the deep,
Through lonely hours their vigil keep.
May peace their mission ever be,
Protect each one we ask of thee.
Bless those at home who wait and pray,
For their return by night or day.

The "submarine verse" is attributed to the late Reverend Gale Williamson who served over twenty years as pastor of the United Methodist Church of Gales Ferry, Connecticut.

---

[18] Lyrics courtesy of Hymnary.org, Calvin University, Grand Rapids, Michigan, used with permission.

What did I learn?

- ✓ "Life is short. We do not have much time to gladden the hearts of those who travel with us. So be swift to love. Make haste to be kind. And go in peace." —Henri-Frédéric Amiel
- ✓ "In ordinary life we hardly realize that we receive a great deal more than we give, and that it is only with gratitude that life becomes rich." —Dietrich Bonhoeffer
- ✓ "Gratitude in its deepest sense means to live life as a gift to be received thankfully. And true gratitude embraces all of life: the good and the bad, the joyful and the painful..."- Henri Nouwen.
- ✓ "Cultivate the habit of being grateful for every good thing that comes to you, and to give thanks continuously. And because all things have contributed to your advancement, you should include all things in your gratitude." —Ralph Waldo Emerson.

Farewell.

May you always enjoy fair winds and following seas.

References

Anderson, Commander William. *Nautilus 90 North*. New York: World Publishing Company, 1059.

Barlow, Jeffrey G. A New Look at the Cuban Missile Crisis. Some Aspects of the U.S. Navy's Participation in the Cuban Missile Crisis. *Colloquium on Contemporary History*. Naval Historical Center. June 18, 1992 No. 7.

Basic Military Requirements, (NAVPERS 10054), Bureau of Naval Personnel, Washington, DC. 1957.

Conrad, Joseph. *Typhoon*. Garden City, New York: Doubleday, Doran & Company, 1942.

Dr. Suess, *Oh, the Places You'll Go*. Random House, 1990.

Drucker, Peter. *The Practice of Management*. Harper and Row, New York. 1954.

London, Jack. *The Sea-Wolf*. 1904. The Reader's Digest Association, Inc. 1989.

Offley, Ed. *Scorpion Down: Sunk by the Soviets, Buried by the Pentagon: The Untold Story of the USS Scorpion*. Basic Books, New York, 2007.

Offley, Ed. The Final Secret of the USS Scorpion. *MHQ—The Quarterly Journal of Military History*. Summer 2018 issue. Vol. 30, No. 4.

Offley, Ed. The Last Secret of the Scorpion. *Navy Times*. May 22, 2019. https://www.navytimes.com/news/your-navy/2019/05/22/the-last-secret-of-the-scorpion

Peters, Thomas J., and Robert H. Waterman, Jr. *In Search of Excellence, Lessons from America's Best-Run Companies*. New York NY: Warner Books, Harper & Row Publishers, 1982.

Peters, Tom. *In Search of Excellence*, Warner Books, Inc. New York, 1984.

Plokhy, Serhii. *Cuban Missile Crisis: Nuclear Folly: A History of The Cuban Missile Crisis*. W.W. Norton, New York, 2021.

Rogers, Dale Evans. *Angel Unaware: A Touching Story of Love and Loss*. Fleming H. Ravell Co, 1953.

Sapolsky, Harvey M. *The Polaris System Development: Bureaucratic and Programmatic Success in Government*, Harvard University Press, 1972.

Scott, James M. *"America's Undersea War on Shipping." Naval History Magazine*. US Naval Institute. December, 2014.

Smith, Hervey Garrett. *The Small-Boat Sailor's Bible*. Garden City, New York: Doubleday & Company, Inc. 1964.

The Bluejackets' Manual, Fifteenth Edition, United States Naval Institute, Annapolis, Maryland. 1959.

The Book of Common Prayer, "Marriage." The Episcopal Church, Oxford University Press, New York, 1990.

ABOUT THE AUTHOR

David Francis is a former nuclear submarine officer and retired US Navy captain with over thirty years of naval service. His business experience includes several years with 3M and IBM, and sales, marketing, and management at five high-technology startup companies in Minnesota. He is the author of *Bringing Jon Home, the Wilderness Search for Jon Francis,* and *Grief Travelers: Learning to Live and Love Again After Loss.*

Francis is executive director of the Jon Francis Foundation (JFF), a nonprofit organization with the mission to provide support and empowerment to the families of people missing in the wilderness. The foundation was created in 2007 to honor the memory and ministry of Jon Francis, David's twenty-four-year-old son, Jon, who went missing in July 2006, while climbing the Grand Mogul in the Sawtooth Mountains of Central Idaho.

David and his wife Linda live in Stillwater, Minnesota. They have been richly blessed with three daughters, Robin, Jocelyn, and Melissa; a son, Jon; and six grandchildren, Taylor, Katie, Stephen, Audrey, Charlie, and Camille.

Contact the author at jonfrancisfoundation.org

Other books by David Francis

Bringing Jon Home: The Wilderness Search for Jon Francis

The National Indie Excellence Award, finalist, Memoir
Midwest Independent Publishers Association, finalist, Inspirational

“Excellent, compelling story. Every parent should read . . . especially fathers.” ~*Midwest Independent Publishers Association.*

“A riveting read.” ~*St. Paul Pioneer Press.*

“An amazing story of love and commitment, honoring the promise that we make to our children: I’ll always be here for you. They did it. They brought Jon home. I wish I had met Jon. This book allows me to meet him and live my life a little deeper because of knowing him. Thank you, David.” ~Patty Wetterling, Jacob Wetterling Resource Center.

“Your book is beautifully structured and written, each word glowing with sorrow and faith and love, a father's true testament of his son.” ~The Reverend Dr. Alla Renee Bozarth, author of *Life is Goodbye, Life is Hello.*

“A real page-turner, David, you've done a remarkable thing for Jon's legacy by not only capturing his giving, involved life, but offering his story as inspiration for other families. As a dad myself, I marvel at your resilience and strength. Your words are driven by the power of your emotion. I can feel it on every page.” ~Kevin Giles, *Minneapolis Star Tribune.*

Grief Travelers: Learning to Live and Love Again After Loss

"*Grief Travelers* by David Francis has the authenticity of one who has been on a long and arduous grief journey himself . . . In service to other families, David Francis has been a saving grace . . ." ~author and poet Alla Renée Bozarth, PhD

"David takes the reader through a personal journey of devastating loss and the many ways in which he and others have persevered in their own lives. *Grief Travelers* will leave those who have suffered ambiguous loss feeling they are not alone . . . The description of what he and others have done to actively grieve provides a pathway others can follow as they go through the confusing, painful, and chaotic process of grieving. It is a book I will recommend to those who struggle with ambiguous loss with a hope that they will find their own way to actively grieve." ~David Carroll, PsyD, LP

"David Francis tells us his story of traveling and carrying grief and what he has learned from it. In the end he shares the final end of his journey—faith, hope, love, and gratitude. I recommend this book to all who experience the pain and grief of our world. May all who read it find hope and grace in their loss." ~The Rev. Dr. Jerry C. Doherty, Episcopal priest and author.

About Red Bike Publishing

Red Bike Publishing provides high quality books which can be found at *www.redbikepublishing.com*, Amazon.com and other fine book stores.

PUBLISHING
Get Rich in a Niche-The Insider's Guide to Self-Publishing in a Specialized Industry

OTHER TOPICS
1. *Rainy Street Stories-Reflections on Secret Wars, Espionage and Terrorism*
2. *2000 Miles On Wisdom*

NOVELS
1. *Commitment-A Novel*
2. *Devoted*

SECURITY BOOKS AND TRAINING

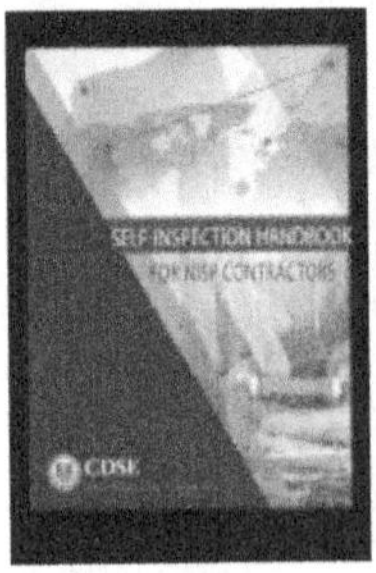